Energy Subsidy Reform

Lessons and Implications

EDITORS

Benedict Clements, David Coady, Stefania Fabrizio, Sanjeev Gupta, Trevor Alleyne, and Carlo Sdralevich

INTERNATIONAL MONETARY FUND

Cataloging-in-Publication Data
Joint Bank-Fund Library

Energy subsidy reform : lessons and implications / editors, Benedict Clements, David Coady, Stefania Fabrizio, Sanjeev Gupta, Trevor Alleyne, and Carlo Sdralevich.— Washington, D.C. : International Monetary Fund, c2013.

       p. : ill. ; cm.

Includes bibliographical references and index.

1. Energy policy—Case studies. 2. Fuel—Prices. 3. Electric utilities—Government policy. I. Clements, Benedict. II. Coady, David. III. Fabrizio, Stefania. IV. Gupta, Sanjeev. V. Alleyne, Trevor. VI. Sdralevich, Carlo. VII. International Monetary Fund.

HD9502.E54 2013

ISBN: 978-1-47555-811-1 (paper)
ISBN: 978-1-47553-252-4 (Mobipocket)
ISBN: 978-1-48439-379-6 (PDF)
ISBN: 978-1-48333-916-9 (ePub)

Publication orders may be placed online, by fax, or through the mail:
International Monetary Fund, Publication Services
P.O. Box 92780, Washington, DC 20090, U.S.A.
Tel.: (202) 623-7430   Fax: (202) 623-7201
E-mail: publications@imf.org
Internet: www.elibrary.imf.org
www.imfbookstore.org

In this work the IMF has changed the debate on energy subsidies. They have argued correctly and persuasively that we must go beyond explicit subsidies and recognize that letting pollution go uncharged is to give a zero price to something that is very costly, in other words, to subsidize. Failing to correct market failure is to undermine the effectiveness of markets. Further, they take the concept forward in serious and important calculations which show just how pervasive energy subsidies are in rich countries and in poor. It is a splendid contribution.

Nicholas Stern
IG Patel Professor of Economics and Government
Chair, Grantham Research Institute on Climate Change and the Environment
London School of Economics
President of the British Academy

This book builds on an extensive cross-country analysis to make recommendations on how to best implement reforms aimed at reducing state subsidies on energy. The first part of the book identifies the costs of not implementing such reforms: (i) an increased burden on public budgets, which reduces the scope for investing more in energy-saving or clean innovations; (ii) an accelerated depletion of natural resources; (iii) an increase in inequality between high- and low-income households; and (iv) more CO_2 emissions, which aggravates the process of global warming. The book also identifies a number of political economy factors to explain governments' reluctance to engage in energy subsidy reforms, in particular, the lack of confidence in governments, and it makes suggestions on how such factors could be (or were) mitigated. The cross-country analysis shows a wide diversity in the way various countries approached the problem. Overall, the book is a gold mine for researchers and policymakers working on energy and sustainable growth and development.

Philippe Aghion
Robert C. Waggoner Professor of Economics, Harvard University

Although our planet has to grapple with the twin problem of scarce energy resources and the adverse environmental consequences of burning fossil fuels, many countries are still subsidizing petroleum products, electricity, natural gas, and coal. This wonderful study offers the best and most comprehensive estimates of such subsidies for a sample of 176 countries. A convincing case is made that these subsidies not only aggravate scarcity and pollution, but also crowd out health and education spending and depress private investment. Furthermore, they are highly distorting and a very inefficient way of redistributing incomes. The 22 country case studies offer warnings, but also suggestions for sustainable reform. This timely book is a must for policymakers in developed and developing countries.

Rick van der Ploeg
Research Director, Centre for the Analysis of Resource Rich Economies
and Professor of Economics, University of Oxford

The best policy-related economics is empirical work that is new, brave, and important. *Energy Subsidy Reform: Lessons and Implications* hits all three targets: new, because the research uncovers the extent and damage of energy subsidies globally, with a scope never before seen; brave, because it challenges both the poorer countries to reconsider what has been a subsidy to the rich cloaked as welfare, and the richer countries to recognize the scale of their own wasteful distortions; important, because it gives practical steps and examples of how this call to action against fiscal and environmental crime can be met. This is the kind of study we need addressing one of the key underappreciated policy disasters of the world today.

Adam Posen
President, Peterson Institute for International Economics

Contents

Acknowledgments

We would like to thank the contributing authors, whose hard work and dedication made this book possible. The book has also benefited from the comments of colleagues from various IMF departments and seminar participants from the Asian Development Bank, the Grantham Institute on Climate Change and the Environment at the London School of Economics, the International Energy Agency, the Organisation for Economic Co-operation and Development, the Petersen Institute, the World Bank, and the World Wildlife Fund. We also thank both the International Energy Agency and the World Bank for providing access to their valuable databases.

We are grateful to Michael Harrup and Sean Culhane of the IMF's Communications Department for managing the editing and production of the book. We also thank Pierre Jean Albert, Jeffrey Pichocki, and Mileva Radisavljevic for their administrative support throughout the entire process. Excellent research assistance was provided by Louis Sears, Kamal Krishna, and Lilla Nemeth.

Benedict Clements
David Coady
Stefania Fabrizio
Sanjeev Gupta
Trevor Alleyne
Carlo Sdralevich

Preface

This volume provides the most comprehensive estimates of worldwide energy subsidies currently available, drawing on data from 176 countries in the areas of petroleum products, natural gas, coal, and electricity. It lays out an analysis of "how to do" energy subsidy reform, drawing on insights from 22 country case studies and analyses carried out by other institutions, and it offers summary narratives and analyses of the reform efforts undertaken in each of those 22 countries. Among the findings that emerged from the analysis are the following.

Energy subsidies have wide-ranging economic consequences. Although aimed at protecting consumers, subsidies aggravate fiscal imbalances, crowd out priority public spending, and depress private investment, including in the energy sector. Subsidies also distort resource allocation by encouraging excessive energy consumption, artificially promoting capital-intensive industries, reducing incentives for investment in renewable energy, and accelerating the depletion of natural resources. Most subsidy benefits are captured by higher-income households, reinforcing inequality. Even future generations are affected through the damaging effects of increased energy consumption on global warming.

Energy subsidies are pervasive and impose substantial fiscal and economic costs in most regions. On a "pretax" basis, subsidies for petroleum products, electricity, natural gas, and coal reached US$492 billion in 2011 (0.7 percent of global GDP or 2 percent of total government revenues). The cost of subsidies is especially acute in oil exporters, accounting for about two-thirds of the total. On a "posttax" basis—which also factors in the negative externalities from energy consumption—subsidies are much higher, at US$2.0 trillion (2.9 percent of global GDP or 8.5 percent of total government revenues). Advanced economies account for about 40 percent of the global posttax total, whereas oil exporters account for about one-third. Removing posttax subsidies could lead to a 15 percent decline in CO_2 emissions and generate positive spillover effects by reducing global energy demand.

Country experiences suggest that there are six key elements for effective subsidy reform. These are (1) a comprehensive energy sector reform plan entailing clear long-term objectives, analysis of the impact of reforms, and consultation with stakeholders; (2) an extensive communications strategy, supported by improvements in transparency, such as the dissemination of information on the magnitude of subsidies and the recording of subsidies in the budget; (3) appropriately phased price increases, which can be sequenced differently across energy products; (4) improvement in the efficiency of state-owned enterprises to reduce producer subsidies; (5) targeted measures to protect the poor; and (6) institutional reforms that depoliticize energy pricing, such as the introduction of automatic pricing mechanisms.

CHAPTER 1

Introduction and Background

TREVOR ALLEYNE, BENEDICT CLEMENTS, DAVID COADY,
STEFANIA FABRIZIO, SANJEEV GUPTA,
AND CARLO SDRALEVICH

The recent surge in international energy prices, combined with incomplete pass-through to domestic prices, has prompted calls to phase out energy subsidies.[1] International energy prices have increased sharply over the past three years, with the exception of natural gas (Figure 1.1). Yet many low- and middle-income economies have been reluctant to adjust their domestic energy prices to reflect these increases. The resulting fiscal costs have been substantial and pose even greater fiscal risks for these countries if international prices continue to increase. In advanced economies, pass-through has been higher, but prices remain below the levels needed to fully capture the negative externalities of energy consumption on the environment, public health, and traffic congestion.

This volume has two principal purposes: first, to review what works best in energy subsidy reform, in light of country experiences globally; and second, to illustrate successes and failures in particular country contexts by summarizing 22 case studies. A novel feature of the volume is that it presents the most comprehensive estimates of energy subsidies available, covering petroleum products, electricity, natural gas, and coal. A central objective of the study underlying this volume was to learn from past subsidy reform experiences, both successful and otherwise, in order to identify key design features that can facilitate reform going forward.

THE GLOBAL SUBSIDIES LANDSCAPE

Any useful discussion of energy subsidies requires some definition of subsidies, particularly because not all subsidies are recorded as expenditures in government budgets. The full spectrum comprises both consumer and producer subsidies. The first type applies to intermediate consumers (firms) and final consumers (households) and the second type to the producers of fuel products, coal, natural gas, and electrical power. Consumer subsidies include two components: a pretax subsidy, which arises if the price paid by firms and households is below supply

---

[1] The G-20 Pittsburgh Communiqué in September 2009 called for a phaseout of inefficient fossil fuel subsidies in all countries. This commitment was reaffirmed at the 2012 Los Cabos meeting of the G-20.

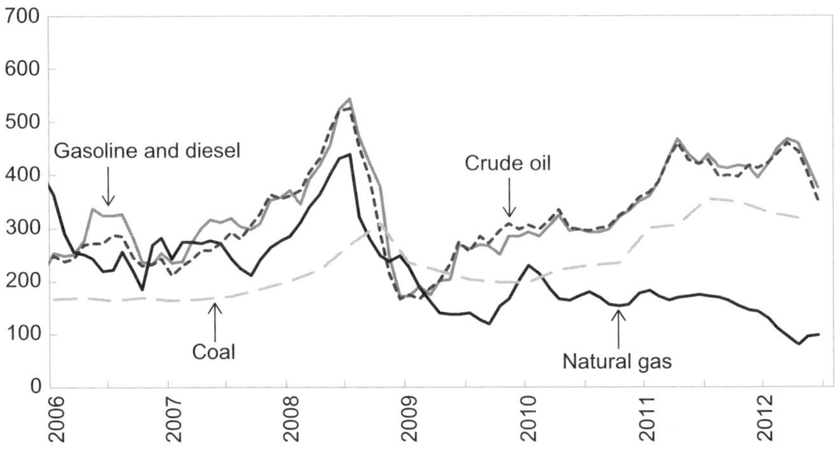

Sources: IMF, *World Economic Outlook* (WEO); Organisation for Economic Co-operation and Development (OECD); U.S. Energy Information Administration (EIA).

Note: Coal price is the average of quarterly U.S. import prices (EIA) and quarterly OECD import price (IEA/OECD). Natural gas price is the average of the monthly U.S. import and export prices (EIA); these prices are weighted averages for liquefied natural gas and pipeline natural gas. Crude oil price is the average of Brent, Dubai, and West Texas Intermediate monthly prices (WEO/Primary Commodities Price System). Gasoline price is the monthly New York Harbor conventional gasoline spot price (EIA). Diesel price is the monthly Los Angeles ultralow sulfur CARB diesel spot price (EIA). Gasoline and diesel prices are then averaged.

Figure 1.1 International Prices of Oil, Coal, and Natural Gas, 2006–12 (Indexed January 2000 = 100)
International energy prices, other than for natural gas, have rebounded since the 2008–9 global crisis.

and distribution costs; and a tax subsidy (if taxes are below their efficient level, which requires that energy products be subject to consumption taxation plus corrective taxes to capture negative environmental and other externalities as a result of energy use, such as global warming and local pollution). These components are further defined in Chapter 2, which also lays out the methods this study used to measure the costs of on- and off-budget energy subsidies in 176 countries.

The study found that pretax subsidies are concentrated in developing and emerging economies, with oil exporters having the largest subsidies. The subsidies' evolution closely follows the trajectory of international energy prices, having declined with the collapse of international prices after the financial crisis but escalating again since 2009. Globally, these subsidies add up to a very large sum, estimated at US$492 billion in 2011, which was over 2 percent of total government revenues.

CONSEQUENCES OF ENERGY SUBSIDIES

Energy subsidies have wide-ranging economic consequences. Subsidy expenditures aggravate fiscal imbalances and crowd out priority public spending and private investment, including in the energy sector. Underpriced energy distorts resource allocation by encouraging excessive energy consumption, artificially promoting capital-intensive industries (thus discouraging employment creation), reducing incentives for investment in renewable energy, and accelerating the depletion of natural resources. Subsidies lead to higher energy consumption, exerting pressure on the balance of payments of net energy importers, while also promoting smuggling to neighbors where domestic prices are higher.

Because most subsidy benefits are captured by higher-income households, energy subsidies have important distributive consequences that are often not fully understood. Even future generations are affected through the reduced availability of key inputs for growth and the damaging effects of increased energy consumption on greenhouse gas emissions and global warming.

Chapter 3 reviews all of these challenges, emphasizing subsidies' fiscal costs, adverse macroeconomic and environmental impacts, and adverse impact on equity on account of their regressive distribution.

CHALLENGES AND SUCCESSES OF REFORM

Energy subsidies have generally been difficult to reform. Subsidy reform has been a frequent topic of discussion between IMF staff and member countries—in some cases over decades. The adjustment of prices for subsidized energy has often led to widespread public protests by those who benefit from subsidies and to either a complete or partial reversal of price increases.[2] The absence of public support for subsidy reform partly reflects a lack of confidence in the ability of governments to reallocate the resulting budgetary savings to benefit the broader population, as well as concerns that vulnerable groups will not be protected.

This is particularly challenging in oil-exporting countries, where subsidies are seen as a mechanism to distribute the benefits of natural resource endowments to their populations; in addition, these countries typically lack capacity to administer targeted social programs. Governments are also often concerned about the inflationary effects of higher domestic energy prices and their adverse impact on the international competitiveness of domestic producers. Furthermore, subsidy reform can be complex when it involves efforts to reduce inefficiencies and production costs, as is often the case for the electricity sector.

Chapter 4 draws on lessons from reform experiences in 22 countries, which cover 28 reform episodes, based on case studies undertaken by IMF staff. These are combined with insights from past IMF analyses[3] as well as from analyses carried out by other institutions.[4] The experiences studied include both successful and

---

[2]Examples of reform reversals where price increases had to be quickly reversed—either partially or fully because of public demonstrations—include Bolivia (2010), Cameroon (2008), Nigeria (2012), Venezuela (1989), and Yemen (2005). Nigeria and Yemen are discussed in this volume among the case studies in Chapters 5 and 7.

[3]Including Gupta and others, 2000; Coady and others, 2006; IMF, 2008a; Coady and others, 2010; and Arze del Granado, Coady, and Gillingham, 2012.

[4]Including Global Subsidies Initiative, 2010; United Nations Environment Program and International Energy Agency, 2002; United Nations Environment Program, 2008; World Bank, 2010b; and Vagliasindi, 2013.

unsuccessful reform episodes over the past two decades across a broad range of countries and different energy products. In some cases, governments attempted to reduce the fiscal burden of subsidies by raising energy prices to households and firms or improving the efficiency of state-owned enterprises. Some governments attempted to reduce pretax subsidies, but others sought to restore energy taxation to higher levels.

The study found cases where countries successfully implemented reforms that led to a permanent and sustained reduction of subsidies (success), others where countries reduced subsidies for at least a year but later saw them reemerge or remain an unresolved policy issue (partial success), and still others where reforms failed, with price increases or efforts to improve efficiency being quickly rolled back (unsuccessful). Out of the 28 reform episodes studied, 12 were classified as a success, 11 as a partial success—often because of reversals or incomplete implementation—and five as unsuccessful.

COUNTRY CASE STUDIES

This volume offers details on the political and economic context and reform efforts in 22 case studies covering petroleum products, electricity, and coal. The selection of countries was designed to ensure coverage of different regions of the world and a mix of reform outcomes. The selection also reflects the availability of data and of previously documented evidence on country-specific reforms. The larger number of studies on fuel subsidies reflects the wider availability of data and past studies of these reforms.

Out of the 22 case studies, 14 address fuel subsidy reform, seven address electricity sector reform, and one addresses coal sector reform. The studies cover seven countries from sub-Saharan Africa (Chapter 5), two countries in emerging and developing Asia (Chapter 6), three countries in the Middle East and North Africa (Chapter 7), four countries in Latin America and the Caribbean (Chapter 8), and three countries in Central and Eastern Europe and the Commonwealth of Independent States (CIS) (Chapter 9). In 14 of the 28 episodes, an IMF-supported program was in place, and in all but two the program contained conditionality on energy subsidy reform.

Defining and Measuring Energy Subsidies

DAVID COADY, STEFANIA FABRIZIO, MUMTAZ HUSSAIN,
BAOPING SHANG, AND YOUNES ZOUHAR

DEFINITION AND MEASUREMENT

Energy subsidies comprise both consumer and producer subsidies. Consumer subsidies arise when the prices paid by consumers, including both firms (intermediate consumption) and households (final consumption), are below supply costs, including transport and distribution costs. Producer subsidies arise when prices are above this level.[1] Where an energy product is internationally traded, such as for petroleum products, the supply cost is based on the international price.[2] For a net importer of fuel products, the supply cost is the overall cost of importing the fuel, whereas for a net exporter, the supply cost represents the forgone revenue, or opportunity cost, from not exporting the product.

Where the energy product is mostly nontraded (such as electricity in most countries), the relevant supply cost is the cost-recovery price for the domestic producer, including a normal return to capital and distribution costs. This approach to measuring consumer subsidies is often referred to as the "price-gap approach" (Koplow, 2009) and is used widely in analyses by other international agencies. In most economies, there are elements of both producer and consumer subsidies, although in practice it may be difficult to separate the two. The advantage of the price-gap approach is that it also helps capture consumer subsidies that are implicit, such as those provided by oil-exporting countries that supply petroleum products to their populations at prices below those prevailing in international markets.

Consumer subsidies include two components: a pretax subsidy, if the price paid by firms and households is below supply and distribution costs, and a tax subsidy, if taxes are below their efficient level. Box 2.1 describes the calculation of these two components. Most economies impose consumption taxes to raise revenue to help finance public expenditures. Efficient taxation requires that all consumption, including that of energy products, be subject to this taxation. The efficient taxation

---

[1] The calculation of producer subsidies should incorporate any subsidies received on inputs.

[2] For some exporters of fuel products with refineries that use domestic crude oil, actual fuel product production costs may be below supply costs if these exporters have access to subsidized crude oil.

Box 2.1   *Pretax and Posttax Consumer Subsidies*

A consumer subsidy is defined as the difference between the supply cost of an energy product, including transport and distribution costs, and the price paid by energy consumers (including both households for final consumption and enterprises for intermediate consumption). There are two concepts of consumer subsidies: pretax subsidies and posttax subsidies.

For the calculation of pretax subsidies for internationally traded goods (such as the refined petroleum products considered in this book), the supply cost is the international price including transport and distribution costs[1] (Pw), so that

$$\text{Pretax subsidy} = Pw - Pc,$$

where Pc is the price paid by consumers. When the good or service is not traded internationally, as is the case for electricity in most countries, then the supply cost is taken as the cost-recovery price (e.g., the costs of generation, transmission, and distribution of electricity). The pretax subsidy is then calculated as above, but Pw is the cost-recovery price. Pretax subsidies exist only in countries where the price paid by consumers is below the supply cost ($Pc < Pw$).

The calculation of posttax subsidies includes an adjustment for efficient taxation ($t^* > 0$) to reflect both revenue needs and a correction for negative consumption externalities:

$$\text{Posttax subsidy} = (Pw + t^*) - Pc,$$

where Pw and Pc are defined as above. Therefore, when there is a pretax subsidy, the posttax subsidy is equal to the efficient tax plus the pretax subsidy. When there is no pretax subsidy, the posttax subsidy is equal to the difference between efficient and actual taxation.

---

[1] When the refined petroleum product is imported, the supply cost is taken as the international fob price plus the cost of transporting the product to the country's border plus the cost of internal distribution. When the product is exported, the supply cost is defined as the revenue forgone by not exporting the product, that is, the international fob price minus the cost of transporting the product abroad (because this cost is saved when the product is consumed domestically rather than exported) plus the cost of internal distribution.

---

of energy further requires corrective taxes to capture negative environmental and other externalities owing to energy use, such as global warming and local pollution.[3] The discussion below focuses on both pretax subsidies and posttax subsidies, where the latter includes an allowance for efficient taxation.

Although energy subsidies do not always appear on the budget, they must ultimately be paid by someone. Whether and how subsidies are reflected in the budget will depend on who incurs them and how they are financed. For example, the cost of pretax consumer subsidies may be incurred by state-owned enterprises

---

[3] These taxes are often referred to as "Pigouvian" or "corrective" taxes. In this paper, only broad estimates of these tax subsidies will be reported. A forthcoming study by the Fiscal Affairs Department of the IMF will provide more refined, country-specific estimates.

(SOEs) that sell electricity or petroleum products at a price below supply costs. If the government fully finances these losses with a transfer, the consumer subsidy will be reflected in the budget as expenditure and financed through higher taxes, increased debt, or higher inflation if the debt is monetized. In many instances, however, the subsidy may be financed by the SOE and reflected in its operating losses or lower profits, lower tax payments to the government, the accumulation of payment arrears to its suppliers, or a combination of all three.

Alternatively, the cost of consumer subsidies could be offset by subsidized access to energy inputs, the cost of which would again fall on the government. In practice, the ways in which subsidies are financed and recorded in the budget vary across countries and can change over time. For example, whereas Indonesia, Jordan, and Malaysia fully record fuel subsidies in the budget, Sudan and Yemen only partially record subsidies, and all subsidies are off budget in Angola. In India, the extent to which fuel subsidies are recorded on budget has varied (Box 2.2). In sum, in one way or another, someone always pays the cost of subsidies.

Pretax Subsidies

Subsidies for petroleum products are calculated for 176 countries using the price-gap approach drawing on data compiled by IMF staff, the Organisation for Economic Co-operation and Development (OECD), and Deutsche Gesellschaft für Internationale Zusammenarbeit (GIZ) for 2000–2011. Consumer subsidies

---

Box 2.2   *Financing Fuel Subsidies in India*

Domestic fuel prices in India have not kept pace with rising international fuel costs, resulting in consumer price subsidies. Reflecting sharp increases in fuel import prices over 2007 and 2008, subsidies peaked at over 2 percent of GDP in fiscal year (FY) 2008/09. As international prices collapsed over the second half of 2008, subsidies also fell sharply to just under 0.9 percent of GDP in FY 2009/10. However, with the rebound in international prices over the last three years, subsidies again started to escalate, reaching nearly 2 percent of GDP in FY 2011/12.

Fuel subsidies have been financed through a number of channels, including off-budget sources. Subsidies are incurred in the first instance by the predominantly state-owned oil marketing companies (OMCs), which sell fuel products to consumers at subsidized prices. These losses incurred by OMCs have been financed in a variety of ways. In FY 2007/08, just under one-half of the financing was recorded on budget, with the remaining half financed off budget. On-budget transfers mainly took the form of so-called government oil bonds issued to OMCs, whereas direct budget transfers to OMCs were negligible. Off-budget financing was split between transfers from state-owned enterprises involved in the upstream production of crude oil and OMCs' self-financing. In effect, OMCs used part of the profits from the sale of other unregulated fuel products to offset these subsidy losses. By FY 2011/12, all on-budget financing took the form of direct budget transfers to OMCs, which accounted for about three-fifths of subsidies, with the remainder financed by upstream transfers.

are estimated for gasoline, diesel, and kerosene. Producer subsidies are included for 12 OECD countries. See Appendix A for details.

Natural gas and coal subsidies are estimated for 56 countries and are largely based on the price-gap approach. These data are mostly drawn from the IEA for 2007–11. Producer subsidies are also included for coal for 16 OECD countries.

A number of different methods are used to estimate electricity subsidies for 77 countries. For some countries in Africa, the Middle East, and emerging Europe, estimates of combined producer and consumer subsidies are compiled from various World Bank and IMF reports. For these countries, subsidy estimates are based on average domestic prices and cost-recovery prices that cover production and investment costs as well as distributional losses and the nonpayment of electricity bills. For other countries, consumer price subsidies are taken from IEA, which derived these values on the basis of the price-gap approach.

Posttax Subsidies

The supply cost was also adjusted for corrective taxes and revenue considerations to estimate posttax subsidies. Rough estimates of corrective taxes, drawing on other studies, were made to account for the effects of energy consumption on global warming; on public health through the adverse effects on local pollution; on traffic congestion and accidents; and on road damage. Estimates of the global warming damages from CO_2 emissions vary widely (see Appendix A). Our estimates assume damages from global warming of US$36 per ton of CO_2 emissions, following the United States Interagency Working Group on Social Cost of Carbon (2013), an extensive and widely reviewed study. For final consumption, posttax subsidy estimates also assume that energy products are subject to the economy's standard consumption tax rate (an ad valorem tax) on top of the corrective tax. The estimates are based on value added tax (VAT) rates for 150 countries in 2011. For countries without a VAT, the average VAT rate of countries in the region with a similar level of income is assumed.

Caveats

These estimates are likely to underestimate energy subsidies and should be interpreted with caution. First, data on producer subsidies are not available for all countries and all products.[4] Second, consumer subsidies for liquefied petroleum gas (LPG) are not included owing to lack of data. Third, fuel subsidy estimates are based on a snapshot of prices paid by firms and households at a point in time (end-of-year) or as an average of end-of-quarter prices when such data are available. Fourth, for electricity, natural gas, and coal, estimates lack full comparability across countries, because they are drawn from different sources and use different approaches. Fifth, they rely on the assumption of similar transportation and distribution margins across countries. Sixth, in light of these factors, our

---

[4]In practice, identifying producer subsidies can be especially difficult because these often take the form of differential tax treatment and tax exemptions for specific sectors.

subsidy estimates may differ from those found in country budget documents, including those reported in the case studies. Seventh, the estimates of corrective taxes are made on the basis of studies for just a few countries and a common assumption regarding how these would vary with country income levels. However, these weaknesses are outweighed by the merits of constructing a broad picture of the magnitude of energy subsidies across as many countries and products as possible.

PRETAX SUBSIDIES

Magnitude of Energy Subsidies

Global pretax energy subsidies are significant. The subsidy estimates capture both those that are explicitly recorded in the budget and those that are implicit and off budget. The evolution of energy subsidies closely mimics that of international energy prices (Figure 2.1). Although subsidies declined with the collapse of international energy prices, they have started to escalate since 2009. In 2011, global pretax subsidies reached US$492 billion (0.7 percent of global GDP or over 2 percent of total government revenues). Petroleum and electricity subsidies accounted for about 45 percent and 30 percent of the total, respectively, with most of the remainder coming from natural gas. Coal subsidies are relatively small at US$6.5 billion.

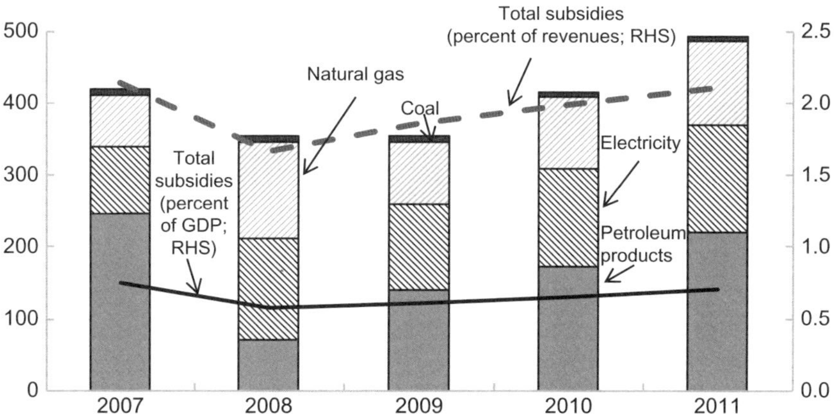

Sources: Deutsche Gesellschaft für Internationale Zusammenarbeit (GIZ); IMF staff estimates; IMF, *World Economic Outlook* (WEO); International Energy Agency (IEA), *World Energy Outlook 2012*; OECD; World Bank.
Note: Data are based on most recent year available. Total subsidies in percent of GDP and revenues are calculated as total identified subsidies divided by global GDP and revenues, respectively.

Figure 2.1   Pretax Energy Subsidies, 2007–11 (Billions of U.S. dollars)
Energy subsidies have surged since the 2008–9 crisis and closely mimic changes in international prices.

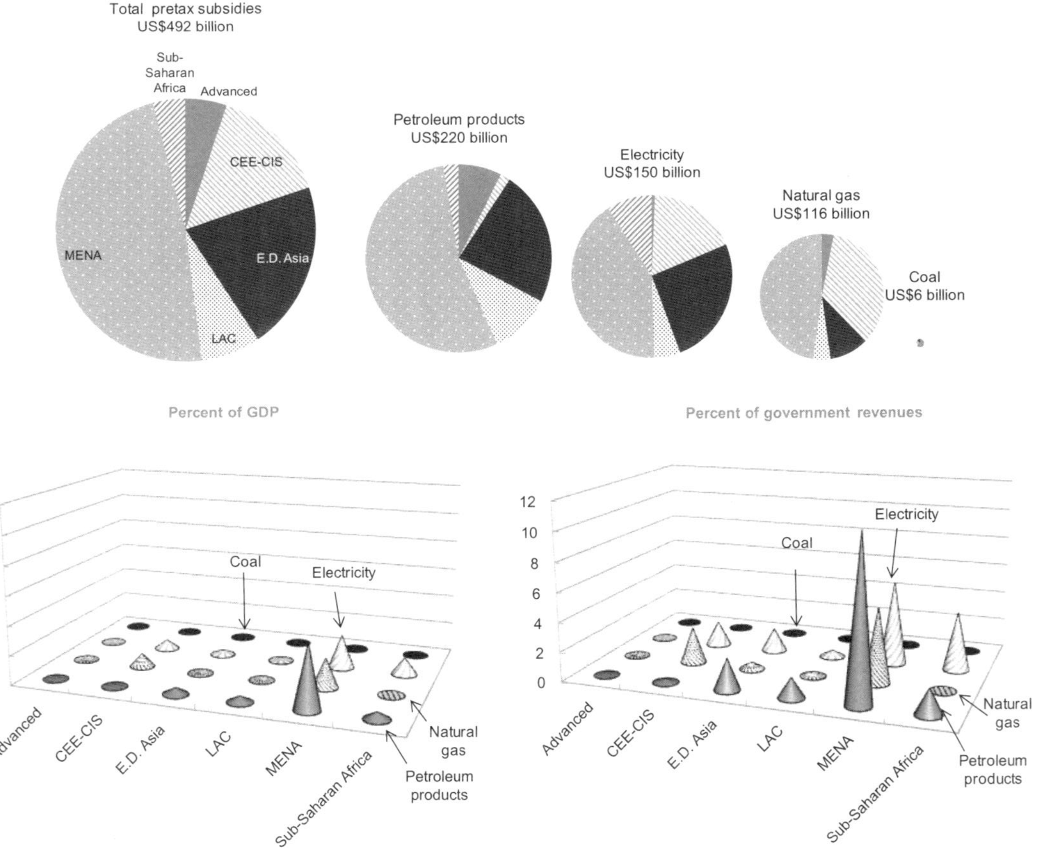

Sources: GIZ; IEA, *World Energy Outlook 2012*; IMF, WEO; IMF staff estimates; OECD; World Bank.
Note: Data for electricity are for the most recent year available. Subsides in percent of GDP and revenues are calculated as identified subsidies divided by regional GDP and revenues, respectively. CEE-CIS = Central and Eastern Europe and Commonwealth of Independent States; E.D. Asia = Emerging and Developing Asia; LAC = Latin America and the Caribbean; MENA = Middle East and North Africa.

Figure 2.2 Pretax Energy Subsidies by Region, 2011
Energy subsidies are concentrated mostly in the Middle East and North Africa, Central and Eastern Europe, and Emerging and

The Geography of Pretax Subsidies

Pretax subsidies are concentrated in developing and emerging economies. Oil exporters—most of which are developing or emerging economies—tend to have the largest subsidies. This finding holds not only when measuring subsidies in absolute terms but also as a share of GDP and on a per capita basis.

The *Middle East and North Africa* region accounted for about 48 percent of global energy subsidies (Figure 2.2, Appendix Table A.2). Energy subsidies totaled over 8.5 percent of regional GDP or 22 percent of total government revenues, with one-half reflecting petroleum product subsidies. The regional average masks significant variation across countries. Of the 20 countries in the region, 12 have energy subsidies of 5 percent of GDP or more. Subsidies are high in this region for both oil exporters and oil importers (Figure 2.3).

Countries in *Emerging and Developing Asia* were responsible for over 20 percent of global energy subsidies. They amounted to nearly 1 percent of regional GDP or 4 percent of total government revenues, with petroleum products and electricity accounting for nearly 90 percent of subsidies. Energy subsidies exceeded 3 percent of GDP in four countries: Bangladesh, Brunei, Indonesia, and Pakistan.

The *Central and Eastern Europe and CIS* countries accounted for about 15 percent of global energy subsidies, including the highest share (at nearly 35 percent) of global natural gas subsidies. Energy subsidies amounted to over 1.5 percent of regional GDP or 4.5 percent of total government revenues, with natural gas and electricity accounting for about 95 percent. They exceeded 5 percent of GDP in four countries: Kyrgyz Republic, Turkmenistan, Ukraine, and Uzbekistan.

Latin America and the Caribbean made up about 7.5 percent of global energy subsidies (approximately 0.5 percent of regional GDP or 2 percent of total government

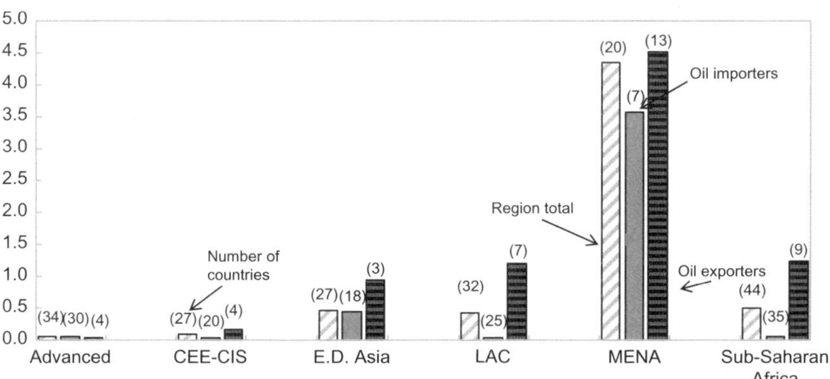

Sources: GIZ; IEA, *World Energy Outlook 2012*; IMF, WEO; IMF staff estimates; OECD; World Bank.
Note: Subsidies in percent of GDP are calculated as identified subsidies divided by regional GDP. Number of countries in each category are indicated in parentheses.

Figure 2.3   Pretax Petroleum Subsidies among Petroleum Importing and Exporting Countries, 2011 (Percent of GDP)
Petroleum product subsidies are systematically higher for oil exporters.

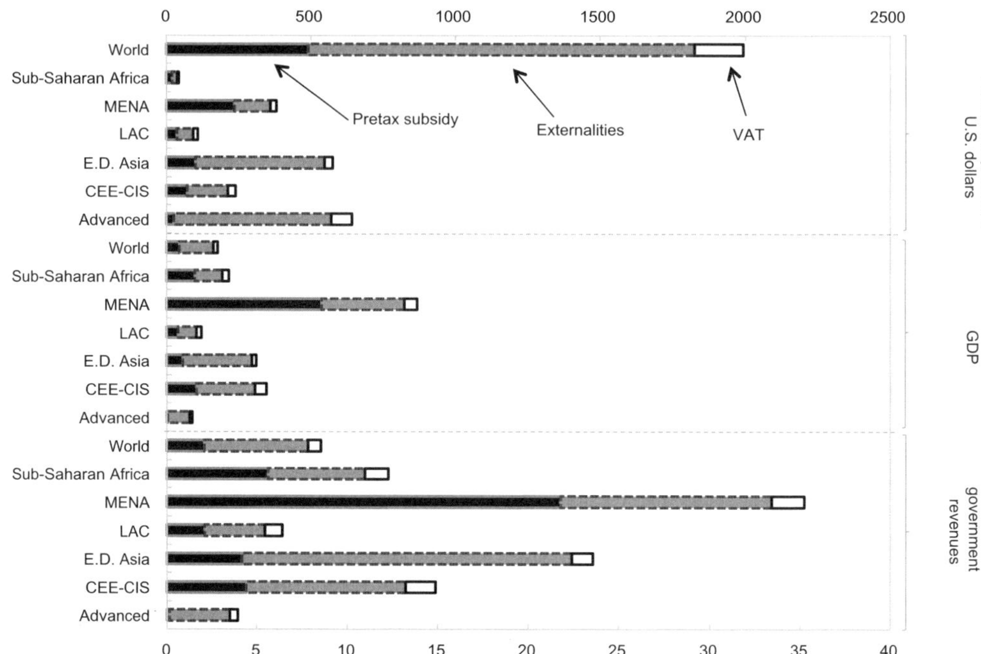

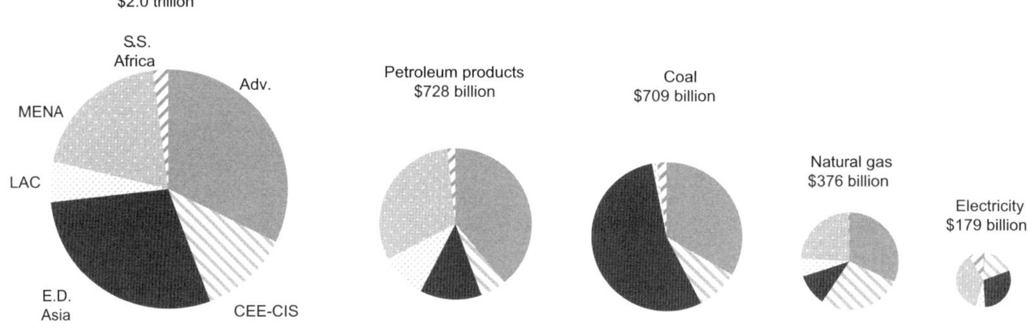

Sources: GIZ; IEA, *World Energy Outlook 2012;* IMF, WEO; IMF staff estimates; OECD; World Bank.
Note: VAT refers to the tax subsidy provided when energy products are taxed by less than the economy's standard VAT rate (see Appendix A). Estimates for electricity are for the most recent year available. Subsidies in percent of GDP and revenues are calculated as identified subsidies divided by global or regional GDP and revenues, respectively.

Figure 2.4 Adjustment of Energy Subsidies for Taxes and Externalities, 2011
Energy subsidies increase substantially when externalities and tax considerations are taken on board.

revenues), with petroleum subsidies accounting for nearly 65 percent. Energy subsidies exceeded 5 percent of GDP in two countries: Ecuador and Venezuela.

Sub-Saharan Africa accounted for about 4 percent of global energy subsidies. Energy subsidies amounted to 1.5 percent of regional GDP or 5.5 percent of total government revenues, with electricity subsidies accounting for over 70 percent. Total subsidies exceeded 4 percent of GDP in three countries: Mozambique, Zambia, and Zimbabwe.

The only advanced economies where energy subsidies were a nonnegligible share of GDP were Belgium at 0.5 percent of GDP (petroleum producer subsidies) and Taiwan Province of China at 0.3 percent of GDP (electricity).

In summary, pretax subsidies are pervasive and impose significant fiscal costs in most developing and emerging regions. They are most prominent in the Middle East and North Africa, especially among oil exporters. Given that energy consumption can be expected to rise as incomes grow, the size of subsidies could climb in regions where they currently account for a small share of the global total, such as sub-Saharan Africa.

POSTTAX SUBSIDIES

By definition, posttax energy subsidies are much larger than pretax subsidies, amounting to US$2.0 trillion in 2011—about 2.9 percent of global GDP or 8.5 percent of total government revenue.[5] Virtually all of the world's economies provide energy subsidies of some kind when measured on a tax-inclusive basis, including 34 advanced economies. For some products, such as coal, posttax subsidies are substantial because prices are far below the levels needed to address negative environmental and health externalities. The fact that energy products are taxed much less than other products also contributes to the high level of posttax subsidies. In the Middle East and North Africa, for example, applying the same rate of VAT or sales taxes to energy products as other goods and services would generate 0.75 percent of GDP. Of the global total, pretax subsidies account for about one-quarter, and tax subsidies account for about three-quarters (Figure 2.4). The advanced economies account for about 40 percent of the global total. The top three subsidizers across the world, in absolute terms, are the United States (US$410 billion), China (US$353 billion), and Russia (US$136 billion).

---

[5]The posttax subsidy figures are calculated by multiplying the subsidy per unit by the quantity of energy consumption in 2011. The revenues gained by eliminating subsidies would be lower than this amount because of the decline in the quantity of energy consumed.

CHAPTER 3

Macroeconomic, Environmental, and Social Implications

ANDREAS BAUER, DAVID COADY, ALVAR KANGUR, CHRISTIAN JOSZ, EDGARDO RUGGIERO, CARLO SDRALEVICH, SUKHWINDER SINGH, AND MAURICIO VILLAFUERTE

HOW SUBSIDIES DEPRESS GROWTH

Energy subsidies depress growth through a number of channels. The effects of subsidies on growth go beyond their adverse impact on fiscal balances and public debt (Kumar and Woo, 2010). Subsidies can discourage investment in the energy sector, crowd out other public spending that would enhance growth, create incentives for smuggling, and over the long term diminish the competitiveness of the private sector.

Subsidies can discourage investment in the energy sector. Low and subsidized prices for energy can result in lower profits or outright losses for producers, making it difficult for state-owned enterprises (SOEs) to expand energy production and unattractive for the private sector to invest in either the short term or the long term (Box 3.1). The result is severe energy shortages that hamper economic activity.[1]

Subsidies can crowd out growth-enhancing public spending. Some countries spend more on energy subsidies than on public health and education (Figure 3.1). Reallocating some of the resources freed by subsidy reform to more productive public spending could help boost growth over the long term.

Subsidies also diminish the competitiveness of the private sector over the longer term. Although in the short term subsidy reform will raise energy prices and increase production costs, over the longer term there will be a reallocation of resources to activities that are less energy and capital intensive and more efficient (Box 3.2), helping spur the growth of employment. Removing energy subsidies helps prolong the availability of nonrenewable energy resources over the long term and strengthens incentives for research and development in energy-saving and

---

[1]Both households and firms spend considerable amounts to address electricity shortages, including through the purchase of generators. For example, in the Republic of Congo, private household and enterprise generator capacity is nearly double the public generation capacity. The cost of own generation by firms is estimated in the range of US$0.3–US$0.7 per kWh—about three to four times as high as the price of electricity from the public grid (Foster and Steinbuks, 2008). These costs are even higher for households because of the smaller generators they use.

> ### Box 3.1 Electricity Subsidies and Growth in Sub-Saharan Africa
>
> *Electricity subsidies in sub-Saharan Africa (SSA) are substantial and primarily reflect high costs of production.* The average cost of subsidized electricity prices in a sample of 30 countries was 1.7 percent of GDP, and in 12 countries it exceeded 2 percent of GDP. On average, the effective tariff in SSA was only about 70 percent of the cost-recovery price during 2005–9. The primary driver of high subsidies has been high costs, rather than low retail prices—residential tariffs in SSA countries are much higher than in other regions of the world. High costs stem from operational inefficiencies, extensive use of backup electricity generation, low economies of scale in generation, and limited regional integration. Therefore, in addition to increasing tariffs, reducing subsidies will require improving operational efficiency and modernizing electricity operations.
>
> *The losses incurred by electricity suppliers owing to subsidized prices have severely constrained their ability to invest in new electricity capacity and improve service quality.* As a result, installed per capita generation capacity in SSA (excluding South Africa) is about one-third of that of South Asia and one-tenth of that in Latin America. Similarly, per capita consumption of electricity in SSA (excluding South Africa) is only 10 kWh per month, compared with roughly 100 kWh in developing countries and 1,000 kWh in high-income countries.
>
> *Deficient electricity infrastructure and shortages dampen economic growth and weaken competitiveness.* Weaknesses in electricity infrastructure are correlated with low levels of productivity (Escribano, Guasch, and Pena, 2008). For example, potential efficiency gains in electricity generation and distribution could reduce costs in the sector by more than 1 percentage point of GDP for at least 18 SSA countries. Simulations based on panel data in Calderón (2008) suggest that if the quantity and quality of electricity infrastructure in all SSA countries were improved to that of a better performer (such as Mauritius), long-term per capita growth rates would be 2 percentage points higher.

alternative technologies. Subsidy reform will crowd in private investment, including in the energy sector, and benefit growth over the longer term.

Finally, subsidies create incentives for smuggling. If domestic prices are substantially lower than those in neighboring countries, there are strong incentives to smuggle products to higher-priced destinations. Illegal trade increases the budgetary cost for the subsidizing country while limiting the ability of the country receiving smuggled items to tax domestic consumption of energy. Fuel smuggling is a widespread problem in many regions around the world, including in North America, North Africa and the Middle East, parts of Asia, and sub-Saharan Africa. For example, Canadians buy cheap fuel in the United States; Algerian fuel is smuggled into Tunisia; Yemeni oil is smuggled into Djibouti; and Nigerian fuel is smuggled into many West African countries (Heggie and Vickers, 1998).[2]

---

[2]In 2011, it was estimated that more than 80 percent of gasoline consumed in Benin was smuggled from Nigeria (IMF, 2012e).

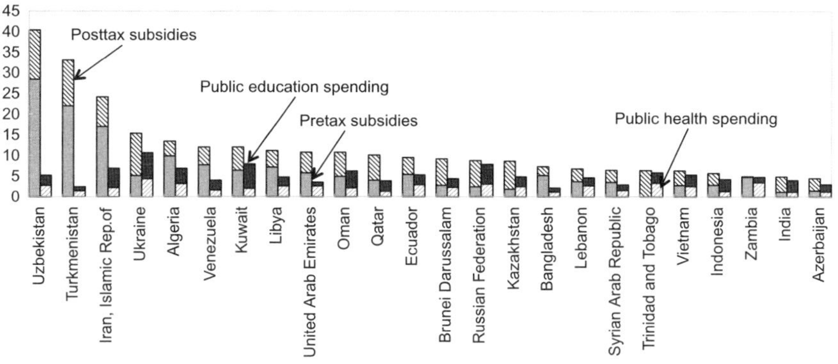

Sources: Clements, Gupta, and Nozaki (2012); IEA, *World Energy Outlook 2012*; IMF staff estimates; OECD; World Bank.
Note: Health and education spending are for 2010 or latest available.

Figure 3.1 Posttax Subsidies and Social Spending, 2010 (Percent of GDP)
Subsidies are substantially higher than critical social spending in many countries.

Box 3.2 *Energy Subsidy Reform and Competitiveness*

The short-term effects of higher energy prices on competitiveness depend on the energy intensity of traded sectors and developments in energy prices in competing countries. Increases in energy prices to reduce subsidies—or to avoid the emergence of subsidies in periods of rising international prices—increase production costs. The effects on costs will vary by sector, depending on both their direct use of energy (e.g., fuel products) and indirect use (e.g., the higher costs of intermediate inputs that use fuel) (Gupta, 1983; Dick and others, 1984). Higher fuel prices, for example, can lead to higher electricity prices, which in turn will affect costs and output in manufacturing (e.g., Clements, Jung, and Gupta, 2007). The use of input–output tables can often be helpful to trace the direct and indirect effects of higher energy prices on costs and competitiveness and to quantify which sectors will be most affected. The effect of higher energy prices on competitiveness depends on developments in energy prices in countries competing for the same markets. If all countries pass on the increase in international prices to domestic prices, for example, the effects on production costs may be similar across countries.

The adverse effects on competitiveness, at the aggregate level, can be reduced if appropriate macroeconomic policies are in place. The extent to which higher energy costs result in a persistently higher price level and adversely affect competitiveness will depend on the strength of "second round" effects on wages and the prices of other inputs (Fofana, Chitiga, and Mabugu, 2009). If prices rise relative to those in trading partners, the real exchange rate will appreciate, reducing competitiveness. These second-round effects can be contained with appropriate monetary and fiscal policies that help anchor inflationary expectations (IMF, 2012d). Subsidy reform helps support an appropriate fiscal policy response by reducing budget deficits and helping contain demand pressures on prices. Flexible exchange regimes also mitigate the impact of volatile international prices on economic growth (IMF, 2008).

> **Box 3.2 (continued)**
>
> *The resources freed from subsidy reform can boost competitiveness over the longer term.* Subsidy reform can contribute to lower budget deficits and interest rates, thus stimulating private investment (Clements, Jung, and Gupta, 2007; Fofana, Chitiga, and Mabugu, 2009). Furthermore, if part of the freed resources is invested in productivity-enhancing public spending, growth dividends can be high (Lofgren, 1995; Breisinger, Engelke, and Ecker, 2011). By removing distortions in price signals, subsidy reform can help reallocate resources toward their best use and improve incentives to adopt energy-saving technologies. Not all sectors will benefit from subsidy reform over the longer term, because those that cannot adapt to higher energy prices will suffer a loss of competitiveness. Yet in the aggregate, the effects on competitiveness are positive. Empirical estimates suggest that higher investment in more efficient and energy-saving technologies could boost growth by up to 1 percent over the long term (von Moltke, McKee, and Morgan, 2004; United Nations Environment Program, 2008; Burniaux and others, 2009; Ellis, 2010).

HOW SUBSIDIES EXACERBATE MACROECONOMIC IMBALANCES

By diluting incentives to reduce domestic energy consumption, the incomplete pass-through to domestic consumers of increases in international energy prices worsens the adverse balance-of-payments impact in oil-importing economies. It also reduces the beneficial balance-of-payments impact in oil-exporting countries. In the latter, the failure to fully adjust domestic prices during periods of rising international prices can make demand management more difficult when higher oil prices boost incomes in the oil sector and lead to higher domestic demand (Gelb and others, 1988). Allowing domestic prices to rise with international prices can help cool off domestic demand during commodity booms and build up fiscal buffers for use during periods of declining prices. To offset concerns about the transmission of high international price volatility to domestic prices, some smoothing of price increases can be considered (see Chapter 4).

OVERCONSUMPTION OF ENERGY AND ITS CONSEQUENCES

The negative externalities from energy subsidies are substantial. Subsidies cause overconsumption of petroleum products, coal, and natural gas and reduce incentives for investment in energy efficiency and renewable energy. This overconsumption in turn aggravates global warming and worsens local pollution. The high levels of vehicle traffic that are encouraged by subsidized fuels also have negative externalities in the form of traffic congestion and higher rates of accidents and road damage. The subsidization of electricity can also have indirect effects on global warming and pollution, but this will depend on the composition of energy sources for electricity generation. The subsidization of diesel promotes the overuse of irrigation pumps, resulting in excessive cultivation of water-intensive crops and depletion of groundwater.

Eliminating energy subsidies would generate substantial environmental and health benefits. To illustrate the impact of subsidies on global warming and local pollution, this study estimated the effects of raising energy prices to levels that would eliminate tax-inclusive subsidies for petroleum products, natural gas, and coal (see Appendix B).[3] The results suggest that this reform would reduce CO_2 emissions by over 5 billion tons, representing a 15 percent decrease in global energy-related CO_2 emissions. Eliminating subsidies would also generate significant health benefits by reducing local pollution from fossil fuels in the form of SO_2 and other pollutants. In particular, this reform would result in a reduction of 12 million tons in SO_2 emissions and a 16 percent reduction in other local pollutants.

The overconsumption of energy products owing to subsidies can also have effects on global energy demand and prices. The multilateral removal of pretax fuel subsidies in non-OECD countries, under a gradual phasing out, would reduce world prices for crude oil, natural gas, and coal by 8 percent, 13 percent, and 1 percent, respectively, in 2050 relative to the no-change baseline (OECD, 2009; IEA, 2011c). The reduction would be substantially larger if prices were raised to levels that eliminated subsidies on a posttax basis. These spillover effects suggest that nonsubsidizers would share the gains from subsidy reform, as well as extending the availability of scarce natural resources.

EQUITY IMPLICATIONS

Energy subsidies are highly inequitable because they mostly benefit upper-income groups. Energy subsidies benefit households both through lower prices for energy used for cooking, heating, lighting, and personal transport and through lower prices for other goods and services that use energy as an input. On average, the richest 20 percent of households in low- and middle-income countries capture six times more in total fuel product subsidies (43 percent) than the poorest 20 percent of households (7 percent) (Figure 3.2).

The distributional effects of subsidies vary markedly by product, with gasoline being the most regressive (i.e., subsidy benefits increase as income increases) and kerosene being progressive. Subsidies to natural gas and electricity have also been found to be badly targeted, with the poorest 20 percent of households receiving 10 percent of natural gas subsidies and 9 percent of electricity subsidies (IEA, 2011a). Although subsidies primarily benefit upper-income groups, a sharp increase in energy prices can nevertheless have a significant impact on the budgets of poor households, both directly through the removal of the subsidies and indirectly through the reduction in real income because of higher consumer prices. For

---

[3]The impact of electricity subsidy removal is not assessed because of data limitations.

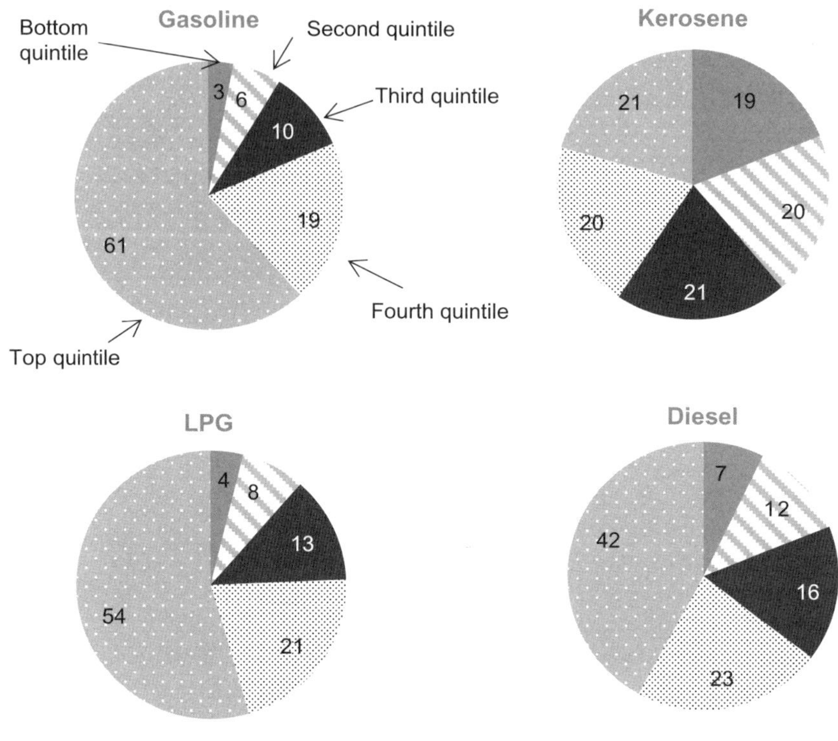

Source: Arze del Granado, Coady, and Gillingham, 2012.
Note: LPG = liquefied petroleum gas.

Figure 3.2 Distribution of Petroleum Product Subsidies by Income Groups (Percent of total product subsidies)
The distribution of subsidies varies across products, with gasoline being the most regressive and kerosene the most progressive.

example, a US$0.25 per liter increase in fuel prices can reduce real consumption of the poorest 20 percent of households by about 5.5 percent (Arze del Granado, Coady, and Gillingham, 2012).

This underscores the need for mitigating measures to ensure that fuel subsidy reform does not result in increased poverty (Sterner, 2012). In the case of electricity, the ability to differentiate tariff levels according to consumption levels (e.g., a lifeline tariff) can help protect low-income groups during electricity subsidy reforms. Nevertheless, such subsidies do not reach poor households that have no access to electricity, which limits the subsidies' progressivity. For example, only 30 percent of households are connected to the grid in sub-Saharan Africa (International Finance Corporation, 2012).

Energy subsidies divert public resources away from spending that is more pro-poor. In many subsidizing countries, equity could be improved by reallocating outlays toward better-targeted programs in health, education, and social protection. Over the longer term, the removal of subsidies, accompanied by a well-designed safety net and an increase in pro-poor spending, could yield significant improvements in the well-being of low-income groups. In oil-exporting countries, subsidies are often used as a tool for sharing oil wealth with citizens. But given the inefficiencies that subsidies create in resource allocation and the high share of benefits that accrues to upper-income groups and in some countries to the expatriate population, energy subsidies are a much less effective policy instrument for distributing wealth than other public spending programs.

CHAPTER 4

Reforming Energy Subsidies: Lessons from Experience

TREVOR ALLEYNE, BENEDICT CLEMENTS, DAVID COADY,
STEFANIA FABRIZIO, SANJEEV GUPTA, CARLO SDRALEVICH,
BAOPING SHANG, AND MAURICIO VILLAFUERTE

OVERVIEW

This chapter provides insights from country case studies to identify ingredients for successful subsidy reform. The country case studies include both successful and unsuccessful subsidy reform episodes over the past two decades across a broad range of countries and different energy products. A total of 22 country case studies were undertaken, covering 28 major reform episodes (Table 4.1). The 22 case studies are presented in detail in Chapters 5 through 9.

The case studies involve episodes in which governments attempted to reduce the fiscal burden of subsidies by raising energy prices to households and firms or improving the efficiency of state-owned enterprises in the energy sector. They contain cases where governments attempted to reduce pretax subsidies but also where governments sought to restore energy taxation to levels that had prevailed prior to increases in international energy prices and to levels needed to eliminate posttax subsidies.[1]

The studies include cases where countries successfully implemented reforms that led to a permanent and sustained reduction of subsidies (success); those that achieved a reduction of subsidies for at least a year but where subsidies have re-emerged or remain a policy issue (partial success); and subsidy reforms that failed, with price increases or efforts to improve efficiency in the energy sector being rolled back soon after the reform began (unsuccessful). Out of the 28 reform episodes, 12 were classified as a success, 11 as a partial success—often because of reversals or incomplete implementation—and five as unsuccessful.

The selection of countries for study was designed to ensure coverage of different regions of the world and a mix of reform outcomes. The selection also reflects the availability of data and of previously documented evidence on country-specific reforms. The larger number of studies on fuel subsidies reflects the wider availability of data and past studies of these reforms. Out of the 22 case studies, 14 address

---

[1]For instance, as a result of subsidy reforms over the late 1980s and the 1990s, Turkey has eliminated subsidies on a posttax basis.

TABLE 4.1

Summary of Country Energy Subsidy Reform Episodes

Region/ Country	Energy product	Reform episode	Reform outcome	Reform impact	IMF-supported program during the reform episode	Conditionality on energy subsidy reform
CEE-CIS						
Turkey	Fuel	1998	Successful	SOEs turned from net loss to net profitability	Yes	Yes
Armenia	Electricity	Mid-1990s	Successful	Electricity sector financial deficit declined from22 percent of GDP in 1994 to zero after 2004	Yes	Yes
Turkey	Electricity	1980s	Successful	Generated additional revenues for maintenance	Yes	Yes
Poland	Coal	1990–1998	Unsuccessful	n.a.	Yes	Yes
	Coal	1998	Successful	The industry became financially viable and achieved substantial reduction in government transfer	No	
Emerging and Developing Asia						
Indonesia	Fuel	1997	Unsuccessful	n.a.	Yes	Yes
	Fuel	2003	Unsuccessful	n.a.	No	
	Fuel	2005	Partially successful	Subsidies declined from 3.5 percent of GDP in 2005 to 1.9 percent in 2006	No	
	Fuel	2008	Partially successful	Subsidies declined from 2.8 percent of GDP in 2008 to 0.8 percent in 2009	No	
Philippines	Fuel	1996	Successful	More than 0.1 percent of GDP	Yes	Yes
Philippines	Electricity	2001	Successful	Subsidies declined from 1.5 percent of GDP in 2004 to zero in 2006	No	
LAC						
Brazil	Fuel	Early 1990s–2001	Successful	From 0.8 percent of GDP in subsidies in mid-1990s to revenue generating since 2002	Yes	Yes

Chile	Fuel	Early 1990s	Successful	n.a.	No	
Peru	Fuel	2010	Partially successful	0.1 percent of GDP	No	
Brazil	Electricity	1993–2003	Successful	0.7 percent of GDP	Yes	Yes
Mexico	Electricity	1999/2001/2002	Unsuccessful	n.a.	Yes	No
MENA						
Iran	Fuel	2010	Partially successful	Growth in the consumption of petroleum products initially stabilized	No	
Mauritania	Fuel	2008	Unsuccessful	n.a.	Yes	No
	Fuel	2011	Partially successful	Subsidies declined from 2 percent of GDP in 2011 to close to zero in 2012	Yes	Yes
Yemen	Fuel	2005	Partially successful	Subsidies declined from 8.7 percent of GDP in 2005 to 8.1 percent in 2006	No	
	Fuel	2010	Partially successful	Subsidies declined from 3.2 percent of GDP in 2010 to 7.4 percent in 2011	Yes	Yes
Sub-Saharan Africa						
Ghana	Fuel	2005	Partially successful	50 percent price increase on average	No	
Namibia	Fuel	1997	Partially successful	More than 0.1 percent of GDP	No	
Niger	Fuel	2011	Partially successful	0.9 percent of GDP	No	
Nigeria	Fuel	2011–12	Partially successful	Subsidies declined from 4.7 percent of GDP in 2011 to 3.6 percent in 2012	No	
South Africa	Fuel	1950s	Successful	Successfully avoided subsidies and secured supply	No	
Kenya	Electricity	Mid-1990s	Successful	Subsidies declined from 1.5 percent of GDP in 2001 to zero in 2008	Yes	Yes
Uganda	Electricity	1999	Successful	2.1 percent of GDP	Yes	Yes

Source: IMF staff.

Note: CEE-CIS = Central and Eastern Europe and Commonwealth of Independent States; LAC = Latin America and the Caribbean; MENA = Middle East and North Africa; n.a.= not applicable; SOEs = state-owned enterprises.

fuel subsidy reform, seven address electricity sector reform, and one addresses coal sector reform. The studies cover seven countries from sub-Saharan Africa, two countries in emerging and developing Asia, three countries in the Middle East and North Africa, four countries in Latin America and the Caribbean, and three countries in Central and Eastern Europe and the Commonwealth of Independent States (CIS). In 14 of the 28 episodes, an IMF-supported program was in place, and in all but two the program contained conditionality on energy subsidy reform.

The findings from the country studies identified in Table 4.1 are complemented with the insights from additional country studies conducted previously by the IMF and others, including Gupta and others (2000); Coady and others (2006); IMF (2008a); Coady and others (2010); Global Subsidies Initiative (2010); UNEP and IEA (2002); UNEP (2008); World Bank (2010b); Arze del Granado, Coady, and Gillingham (2012); and Vagliasindi (2013).[2] They also draw on lessons from technical assistance reports on energy subsidies undertaken by the Fiscal Affairs Department of the IMF.[3]

BARRIERS TO REFORM

Country reform experiences suggest a number of barriers to successful subsidy reform. Although there is no single recipe for success, addressing these barriers, which vary from country to country, can increase the likelihood of reforms achieving their objectives and help avoid policy reversals.

Lack of Information Regarding Subsidies

The full fiscal cost of energy subsidies—including both producer and consumer subsidies—is rarely reflected in the budget. This is especially the case for oil exporters, because the subsidies provided by low energy prices are often implicit, that is, not explicitly recorded in the budget.[4] Populations are also often unaware of how domestic energy prices compare with international market prices, the consequences of low energy prices for both the budget and economic efficiency, and the benefit distribution of energy subsidies. As a result, the public is unable to make a connection between subsidies, constraints on expanding high-priority public spending, and the adverse effects of subsidies on economic growth and poverty reduction. This is especially important for oil exporters, where subsidies are very large.

---

[2]The case studies do not disentangle the effects of subsidy reform on macroeconomic variables, such as inflation and the real exchange rate. This would require isolating these effects over the period in which subsidy reforms were implemented, which on average was five years.

[3]Over the past five years, there were 19 technical assistance missions to member countries addressing the issue of energy subsidy reform. About one-third of these missions were to sub-Saharan Africa and another third were to the MENA region.

[4]Gupta and others (2004) estimate implicit subsidies in oil exporters at 3.5 percent of GDP, on average, in 1999.

Out of the 28 reform episodes, 17 indicate that lack of information was a barrier to reform, including fuel subsidy reforms in Ghana, Mexico, Nigeria, the Philippines, Uganda, and Yemen. Yemen's experience is instructive, in that the public accepted a large adjustment in fuel prices when it had been made aware of the need for reforms and of their benefits, but when reforms were introduced without an effective public information strategy, especially during political crises, popular protests forced at least a partial reversal of the adjustments.

Lack of adequate communication to the public was a barrier in electricity subsidy reforms in Mexico and Uganda. In Mexico, the failure of reform efforts in 1999 stemmed from a variety of challenges, but given the public's generally negative attitude toward privatization and its limited awareness of the problems in the electricity sector, the government's failure to roll out a comprehensive communications program was a major contributor.

Most countries that successfully reformed energy subsidies undertook an evaluation of the magnitude of energy subsidies prior to implementing subsidy reforms. Public discussions based on such studies were an important component of the information campaigns in fuel subsidy reforms in Ghana, Namibia, and the Philippines. The Philippines' success in fuel price reform in the 1990s benefited from an exceptionally well-designed communication strategy, which focused on consensus building. Initially, the political environment was not conducive to the proposed reform, which lacked support from the majority party in both legislative chambers, but the administration quickly launched a nationwide road show to inform the public of the problems of oil price subsidies. It also set up a coordination body between the executive and congress and used it to forge a political consensus.

Lack of Confidence in the Government

Even where the public recognizes the magnitude and shortcomings of energy subsidies, it often has little confidence that the government will use savings from subsidy reform wisely. This is especially true in countries with a history of widespread corruption, lack of transparency in the conduct of public policy, and perceived inefficiencies in public spending. The middle class may fiercely resist the removal of these subsidies because they are viewed as one of the few concrete benefits they receive from the state. This is especially the case for oil exporters that have ample fiscal resources yet lack the administrative capacity to implement cash transfer programs.

Lack of credibility was seen as an important factor behind the less successful fuel subsidy reforms in Indonesia in 2003 and Nigeria in 2011. Indonesia's 2003 effort to automatically link movements in domestic fuel product prices to international prices failed in part because it was poorly communicated to a public that already distrusted its government. Protesters there believed that powerful interest groups would benefit, and the public was generally opposed on the grounds that political corruption and inefficiency would undermine reform. In addition, many of the announced compensation programs did not materialize. As a result, the government rolled back most of that year's reform.

Concerns over Harmful Impact on the Poor

Although most of the benefits from energy subsidies are captured by higher-income groups, as noted earlier, energy price increases can still have a substantial adverse impact on the real incomes of the poor, both through higher energy costs of cooking, heating, lighting, and personal transport and through higher prices for other goods and services, including food. This is an important consideration for countries that do not have a well-functioning social safety net capable of effectively protecting the poor from the adverse impact of higher energy prices. In 20 reform episodes, subsidy reform was accompanied by specific measures to mitigate the impact of price increases on the poor. In seven episodes, price increases were initially concentrated on products that were less important for poor household budgets.

Concerns over General Economic Impact

Other concerns include a potential adverse impact on inflation and on international competitiveness, as well as on the volatility of domestic energy prices. Increases in energy prices will have short-term effects on inflation, which may give rise to expectations of further increases in prices and wages unless appropriate macroeconomic policies are in place (Box 3.2). This may especially be a concern for countries that have difficulty in anchoring inflation expectations. Higher energy prices may also lead to concerns about the international competitiveness of energy-intensive sectors. In addition, countries are hesitant to liberalize energy prices in order to avoid high volatility in domestic prices arising from international price developments.

In Armenia, the impact of electricity price increases on inflation was mitigated by the implementation of macroeconomic stabilization measures. In Iran and Nigeria, fuel subsidy reform was accompanied by specific measures intended to mitigate the impact of price increases on energy-intensive sectors. In its reform beginning in 2011, the Nigerian government promised to use savings from the reduction in fuel subsidies to put in place programs to stimulate the economy, including through critical infrastructure projects in the power, roads, transportation, water, and downstream petroleum sectors.

Opposition from Interest Groups

Opposition may arise from specific interest groups benefiting from the status quo. Politically vocal groups that benefit from subsidies can be powerful and well organized and can block reforms. For example, in some countries, the urban middle class and the industrial sector (which also benefits from subsidies) can be obstacles to reform. Conversely, those benefiting from reform are often dispersed and less organized. Reform strategies therefore need to address the concerns of the losers.

An important stumbling block to reform in many countries is often state-owned enterprises (SOEs) in the energy sector, which can resist efforts to strengthen governance and performance. In Mexico, the electricity sector is dominated by the government-owned Comisión Federal de Electricidad, which even after deregulation of power generation in 1992 still accounts for three-quarters of total generation capacity and monopolizes transmission and distribution. Moreover, this dominance of the public sector is mandated by constitutional provisions.

Strong opposition from labor unions can also contribute to the failure of reforms. This was true in Mexico as well as in Poland. In Poland, initial mining sector reforms were unsuccessful because they did not provide adequate support for miners, the people most directly and negatively affected by the reforms, who had a strong lobby. Mitigating measures designed in cooperation with the unions and included in subsequent reform plans broke the resistance of the miners to the restructuring. The experience demonstrated that unions have an important role in the reform process. This is especially likely to be true for reforms affecting any industry that is both a dominant employer within the economy and one whose employees have very specialized skills of limited use outside the industry.

Weak Macroeconomic Conditions

Public resistance to subsidy reform is lower when economic growth is relatively high and inflation is low—although subsidy reform cannot always be postponed and is often required as part of efforts to constrain inflation and stimulate growth. Rising household incomes can help households better afford the increases in energy prices entailed by subsidy reform. In Peru, the implementation of subsidy reforms in early 2010 during a period of stable prices and strong economic growth helped make the reform more politically palatable. In Turkey, reforms of the electricity sector coincided with a period of economic growth and improving standards of living, which assured the public that reforms were moving the country in the right direction.

High inflation is also an obstacle to reform. When inflation is high, frequent large changes in controlled prices are needed to avoid the emergence of fuel subsidies. In Brazil, high rates of inflation and currency depreciation during the 1990s made containing the fiscal costs of subsidies a difficult challenge. To avoid the emergence of subsidies in Brazil, frequent price increases were necessary. This succeeded for other fuels but failed in the case of diesel, whose price increases did not keep pace with exchange-rate depreciation, leading to an upward spike in diesel subsidies.

DESIGNING A SUBSIDY REFORM STRATEGY

Many countries have incorporated specific measures into their subsidy reform strategies to overcome the above barriers. Our review of country reform experiences suggests that the following key elements can increase the likelihood of successful subsidy reform: (1) a comprehensive reform plan; (2) a far-reaching communications strategy, aided by improvements in transparency; (3) appropriately phased energy price increases, which can be sequenced differently across energy products; (4) improvement in the efficiency of SOEs to reduce producer subsidies; (5) targeted mitigating measures to protect the poor; and (6) depoliticization of energy pricing to avoid the recurrence of subsidies. Each of these elements is discussed in turn below in more detail.

The Reform Plan Should Be Comprehensive

Most of the successful reforms were well planned and based on a clear reform strategy. In Iran, the 2010 fuel subsidy reform incorporated clear objectives, compensating measures, and a timetable for reform, preceded by an extensive public relations campaign. The public information campaign emphasized that the main objective of the reform was to replace price subsidies with cash transfers to reduce incentives for excessive energy consumption and smuggling. Bank accounts were opened for most citizens prior to the reform, and compensating cash transfers were deposited into these accounts preceding the implementation of price increases.

In Namibia, the authorities undertook comprehensive planning, including broad consultation with civil society and a well-crafted plan that involved the introduction of a fuel price adjustment mechanism and a targeted subsidy for those living in remote areas. A clear medium-term reform strategy backed by careful planning was also a major factor behind the successful electricity price liberalization reforms in the Philippines and Turkey. By contrast, the lack of effective planning contributed to less successful outcomes in some countries (fuel subsidy reform in Indonesia in 1998 and only partial success in Nigeria in 2011). A good reform plan often requires extensive time to prepare, as it did in Iran.

A comprehensive reform plan requires (1) establishing clear long-term objectives, (2) assessing the likely impact of reforms, and (3) consulting with stakeholders.

Clear Long-Term Objectives

Subsidy reforms are more likely to be successful and durable if they are embedded within a broader reform agenda. In particular, reforms should incorporate both a sustainable approach to energy pricing and a plan to improve the efficiency of energy consumption and supply.

In the Philippines and Turkey, full price liberalization and structural reform of the energy sector, for both fuel and electricity, were articulated as the ultimate goals of reform. This contributed to the eventual success of reform because the public and governments were able to focus on and adhere to long-term goals, without being distracted by setbacks at intermediate stages.

This comprehensive strategy is especially important for electricity reforms. There is a strong inverse correlation between the size of electricity subsidies and the quality of service, reflecting the dampening effect of subsidies on investment. Yet the public is often unwilling to pay higher prices in the absence of quality improvements. Reforms in this sector should not only seek to improve access and service quality but also tackle operational inefficiencies (such as high distribution losses and inadequate bill collection and metering). The need to accompany tariff increases with service improvements can constrain the speed of reform, because improving services often requires up-front investment.

Electricity subsidy reforms in Armenia, Brazil, and Kenya were successful because they were part of a broader package intended to address supply problems. In Armenia, electricity tariff reforms were complemented by institutional reforms, paving the way for private-sector participation, which in turn brought gains in efficiency. Losses in the power supply system declined from 30 percent to 10 percent in 11 years' time.

Assessment of Likely Reform Impact

Designing a comprehensive subsidy reform strategy requires information on the likely impact of reforms on various stakeholders and the identification of measures to mitigate adverse impacts. This involves assessing the fiscal and macroeconomic effects of subsidies and identifying the winners and losers from reform. In Ghana, in 2005, the government commissioned an independent poverty and social impact analysis to assess the winners and losers from fuel subsidies and subsidy removal. This was an important foundation for persuasively communicating the necessity for reform and for designing policies to reduce the impact of higher fuel prices on the poor. In Nigeria, by contrast, the National Assembly did not support the removal of the gasoline subsidy in 2011, claiming a lack of firm data underpinning the size and incidence of subsidies.

Consultation with Stakeholders

Stakeholders should be invited to participate in the formulation of the subsidy reform strategy. This stakeholder approach has proven successful in a number of countries (Graham, 1998; Gupta and others, 2000). In Kenya, electricity tariff increases faced significant difficulties early in the reform process. These were overcome after intense negotiations with stakeholders, particularly with large consumers, and efforts to communicate the objectives and benefits of the reform.

In Namibia, the National Energy Council, chaired by the Minister of Mines and Energy, established a National Deregulation Task Force to examine fuel price deregulation through a consultative process. The task force recommended keeping targeted subsidies to remote areas, deregulating prices gradually, and enhancing transparency in the handling of government fuel tax revenues. Price smoothing, complemented by mitigating measures such as pump price subsidies for rural fuel stations, were key to the reform's eventual success over the following decade. Similarly, in Niger, the authorities established the Comité du Différé to discuss the best way to approach the fuel subsidy reforms and their subsequent consultation with all relevant stakeholders.

By contrast, in Indonesia, consultation with stakeholders had been inadequate in the run-up to the failed 2003 fuel subsidy reform. The widespread and sometimes violent opposition to that reform, as mentioned earlier, was partly motivated by the belief that the reform favored powerful interest groups. The partial success of Indonesia's 2005 reform, as well as the reduced intensity of protests against it, has been credited by some to the government's decision to compensate poor households for the increase in their living costs by establishing welfare programs.

Political and Public Support Require a Communications Strategy and Transparency

Key Messages

A far-reaching communications campaign can help generate broad political and public support and should be undertaken throughout the reform process. A review of subsidy reform experiences found that the likelihood of success almost tripled with strong public support and proactive public communications (IMF, 2011b). The information campaign should explain the magnitude of energy subsidies and their implications for other parts of the budget. The benefits of removing subsidies, including on a posttax basis, should be underscored, in particular the scope for using part of the budgetary savings or additional revenues to finance high-priority spending on education, health, infrastructure, and social protection.

Information campaigns have underpinned the success of a number of countries, including fuel subsidy reforms in Ghana, Iran, Namibia, and the Philippines and electricity subsidy reforms in Armenia and Uganda.

In the Philippines, as discussed earlier, the public communication campaign began at an early stage and included a nationwide road show to inform the public of the problems of petroleum price subsidies. In Namibia, a white paper on energy policy formed the basis of an effective public communications campaign. In Uganda, the government effectively communicated the cost of the electricity subsidy and its incidence to the public. The Ugandan government argued that it lacked the resources to continue subsidizing electricity for the relatively rich elite. Because 88 percent of the population lacked any access to electricity, the limited protests in Kampala over rate increases gained little sympathy. A large portion of the media as well were persuaded by the government's facts and editorialized that raising the tariffs would be a pro-poor measure.

Transparency

Ensuring transparency is a key component of a successful communications strategy. Useful information to be disseminated includes

1. the magnitude of subsidies and how they are funded, including in oil-exporting countries where subsidies are provided implicitly and not shown in the budget or recorded as tax expenditures. To the extent that subsidies are off budget, they could be reported as a memorandum item in budget documents. Data on subsidies should also cover producer subsidies, which may necessitate better reporting of the accounts of SOEs in the energy sector and reporting on SOEs in budget documents;

2. the distribution of subsidy benefits across income groups;

3. changes in subsidy spending over time; and

4. the potential environmental and health benefits from subsidy reform.

Prior to its successful subsidy reform, Niger started to record fuel subsidies explicitly in the budget. Making such information public allows for an independent assessment of the costs and benefits of subsidy policies. It is particularly important for determining whether subsidies are the most effective way to achieve desired outcomes, such as social protection for the poor. Subsidy expenditures should be compared with spending on priority areas and planned increases in these outlays as a consequence of the enlarged fiscal space from subsidy reform. Governments should also disclose as much information as possible about how prices are formulated and the factors behind planned price increases. Both Ghana and South Africa regularly publish such details for petroleum products on their government Web sites and in the national media.

Price Increases Need to Be Appropriately Phased and Sequenced

Phasing in price increases and sequencing them differently across energy products may be desirable. The appropriate phasing in and sequencing of price increases will depend on a range of factors, including the magnitude of the price increases required to eliminate subsidies, the fiscal position, the political and social context in which reforms are being undertaken, and the time needed to develop an effective communications strategy and social safety nets. In the case studies, successful and partially successful subsidy reforms required, on average, about five years.

Pace and Timing of Price Increases

Too sharp an increase in energy prices can generate intense opposition to reforms, as happened with fuel subsidy reforms in Mauritania in 2008 and Nigeria in 2012. A phased approach to reforms permits both households and enterprises time to adjust and permits the country time to build credibility by showing that subsidy savings are being put to good use. As noted earlier, it also helps reduce the impact of subsidy reform on inflation and creates room for governments to establish supporting social safety nets.

The case studies show that 17 out of the 23 reform episodes that were successful or partially successful involved a phased reduction of subsidies. In Namibia, subsidies were removed steadily according to a three-year reform plan. In Brazil, the government pursued a step-by-step approach to reforming petroleum subsidies during the 1990s in order to minimize opposition from key interest groups. Despite initial sharp increases in prices, gradual adjustment of fuel prices was a key design feature of the reforms introduced in Iran, where the plan was to eliminate petroleum subsidies over a five-year period. A gradual approach was also adopted by Kenya (electricity), where the authorities were able to progressively gain support for broader reform by delivering improved services.

The timing of energy price increases should also be considered carefully. For example, coordinating increases in electricity tariffs with the expansion of capacity, as in Uganda, could help win broad acceptance. Tariff increases that coincide with price increases for other socially sensitive products, such as food and fuels, may meet strong resistance.

Sequencing of Reform

Price increases can also be sequenced differently across energy products. For example, petroleum price increases can initially be larger for products that are consumed more by higher-income groups and by industry, such as gasoline and jet kerosene. As the safety net is strengthened, subsequent rounds of reform can include larger increases in prices for fuel products that are more important in the budgets of poor households, and part of the budgetary savings can be used to finance targeted transfers to poor households. For electricity, tariff increases can initially focus on large residential users and commercial users. Out of the 28 reform episodes, seven reforms sequenced price increases in this way.

In Brazil, for instance, petroleum product reforms started by liberalizing prices for products used primarily by industry, followed by a more extensive liberalization of gasoline prices and, finally, of diesel prices. Reforms in Peru initially focused on lifting the subsidy of high-octane gasoline, which is used by luxury cars, allowing international price changes to be fully passed on to domestic prices. A year later, in 2012, the subsidy of regular gasoline was also removed. Peru's reform has been successful in reducing the fiscal cost of the subsidy without provoking widespread opposition. At the same time, it never touched the most politically sensitive products, diesel and liquid petroleum gas (LPG), which also represent the largest share of subsidy spending.

Challenges Created by Gradual Reform

First, a slower pace of reform reduces budgetary savings in the short term. There is thus a trade-off between the objectives of achieving budgetary savings and softening the impact of reforms on households. Second, sequencing of reforms can severely distort consumption patterns. For example, there is a limit to how low kerosene prices can be maintained without serious disruption of energy markets when other petroleum product prices are raised. These problems include the redirection of kerosene and LPG from households to the transport sector and cross-border smuggling. Turkey had to curtail LPG subsidies more rapidly than planned because of a sharp increase in LPG consumption as a result of the conversion of vehicles to LPG. Third, a gradual reform runs the risk that opposition may build up over time.

To address these concerns, gradual reforms must be accompanied by the government's long-term commitment to follow through on planned price increases, possibly over several successive administrations. This challenge can be overcome by building up a broad support base for reforms. For example, Turkey started toward a more liberalized regime for energy pricing, including fuel and electricity, in the late 1980s and early 1990s and continued implementing the plan under subsequent administrations. Effective planning and communication promoted broad consensus on the need for petroleum and electricity sector reforms in the Philippines and enabled the government to successfully implement its reform strategy gradually.

The Efficiency of SOEs Needs to Be Improved

Improving the efficiency of SOEs can reduce the fiscal burden of the energy sector. Energy producers often receive substantial budgetary resources—in terms of both current and capital transfers—to compensate for inefficiencies in production and revenue collection. Improvements in efficiency can strengthen the financial position of these enterprises and reduce the need for such transfers. Country experiences suggest the importance of strengthening SOE governance, improving demand management and revenue collection, and better exploiting scale economies to improve enterprise efficiency.

Governance

Governance of SOEs can be strengthened by improving the reporting of information on operations and costs. This can help identify system inefficiencies (e.g., overstaffing) and vulnerabilities (e.g., major loss points and bottlenecks in energy flows). Countries that have adopted information systems include Kenya, Uganda, and Zambia. Consistent with the Code of Good Practices on Fiscal Transparency, all extrabudgetary activity of the central government, including that undertaken by SOEs, should be reported in budget documents (see also IMF, 2012a).

A second step is to set performance targets and incentives on the basis of this information. In Cape Verde, the electricity company is allowed to keep resources from overperformance on its targets, which can then be used for investment. Introducing competition, including from the private sector, can strengthen performance. This option will be more viable for countries with larger markets, where there is scope to "unbundle" activities in both the petroleum and electricity sectors. Notwithstanding these limitations, the private sector's role in the electricity sector is growing in many emerging and low-income countries. Many of these countries have permitted competition among private generation companies, and some of them have invited the private sector to manage electricity distribution, primarily to address operational inefficiencies.

Demand Management

Improved demand management (by charging higher prices during peak periods) has proven effective in shifting demand to periods where marginal costs of provision are lower (Antmann, 2009). Utilities in sub-Saharan Africa have had programs to provide free compact fluorescent bulbs, which have helped reduce demand and costs in Cape Verde, Ethiopia, Malawi, Uganda, and Rwanda. Revenue-enhancing measures include improved collection and metering. These efforts can start with large customers and then gradually extend to medium-size and smaller ones.

Regional Integration

Efficiency can be improved by exploiting regional trade in electricity (Foster and Briceño-Garmendia, 2010). For instance, Mali and Burkina Faso have been able to expand domestic supply and household access through integration into the regional market.

Well-Targeted Measures Are Needed to Mitigate Impact on the Poor

Well-targeted measures to mitigate the impact of energy price increases on the poor are critical for building public support for subsidy reforms. The first step in this regard is to assess the capacity to expand existing (or implement new) social programs in the short term. Implementing or expanding targeted programs immediately prior to price reforms can help demonstrate the government's commitment to protecting the poor. Untargeted cash transfers to compensate the population following a subsidy reform could be limited to the amount consumed by the poorest. This can generate fiscal savings, because poor households typically consume substantially lower quantities of energy than the rich.

Further fiscal savings would be generated by targeted cash transfers to compensate only lower-income groups. In some oil-exporting countries, where subsidies are often seen as a form of wealth sharing, uniform per capita transfers can be both more efficient and more equitable than untargeted energy subsidies. However, wealth sharing may be better achieved through targeted and productive public spending aimed at building physical and human capital. The degree to which compensation should be targeted is a strategic decision that involves trade-offs between fiscal savings, capacity to target, and the need to achieve broad acceptance of the reform. Out of the 28 reform episodes, 18 relied on targeted mitigating measures, including expansion of public works, education, and health programs in poor areas.

Targeted Cash Transfers

Targeted cash transfers or near-cash transfers (vouchers) are the preferred approach to compensation. Cash transfers give beneficiaries the flexibility to purchase the level and type of energy that best suits their needs and at a time and place of their choosing. They also remove the need for governments to be directly involved in the distribution of subsidized energy to households, which is often extremely costly and prone to abuse (Grosh and others, 2008). Targeted cash transfers were used to protect poor households in nine out of the 28 reform episodes.

Indonesia's unconditional cash transfer program, which covered 35 percent of the population, was an important component of its successful strategy in overcoming social and political opposition to fuel subsidy reforms. Its experience also suggests that such programs need good preparation and monitoring in order to effectively assist the poor.

In Armenia, the government accompanied its electricity subsidy reforms with a series of social safety net reforms. These included a targeted cash transfer program, the Poverty Family Benefit, which has helped beneficiaries maintain real consumption in the face of higher electricity bills. The benefit's design has also helped increase the collection rate and improved energy efficiency, because it is withdrawn from households that overconsume or do not pay their electricity bills. It initially covered 25 percent of households in the country, but coverage gradually declined to 18 percent (as of 2010) as eligibility criteria were tightened, a measure that allowed the average benefit payment to rise by 40 percent in real terms while maintaining the program's overall cost at around 1 percent of GDP.

In addition, two one-off cash transfers were made to low-income households in 1999–2000 to help them cope with higher electricity prices. Beneficiaries included eligible households under the Poverty Family Benefit program and other households considered to have difficulties paying their bills.

The recent expansion of conditional cash transfer programs throughout emerging and low-income economies, with eligibility for benefits linked to household investments in the education and health status of family members, has greatly increased the capacity of these economies to protect poor households from price and other shocks while simultaneously addressing the root causes of persistent poverty (Fiszbein and Schady, 2009; Garcia and Moore, 2012).

Other Programs

When cash transfers are not feasible, other programs can be expanded while administrative capacity is developed. They should focus on existing programs that can be expanded quickly, possibly with some improvements in targeting effectiveness, such as school meals, public works, reductions in education and health user fees, subsidized mass urban transport, and subsidies for consumption of water and electricity below a specified threshold. This approach was used in 15 of the reform episodes, sometimes in conjunction with targeted cash transfers.

In the context of fuel subsidy reforms, targeted social spending programs were expanded to protect lower-income households from fuel price increases in Gabon, Ghana, Niger, Nigeria, and Mozambique. In Ghana, measures included the elimination of fees for state-run primary and secondary schools, a price ceiling on public-transport fares, increases in the minimum wage, purchases of additional public-transport buses, and funding for health care in poor areas. Ghana also increased its investment in electrification in rural areas. The Philippines maintained college scholarships for low-income students and subsidized loans to enable engines used in public transportation to be converted to less costly LPG; it also maintained electricity subsidies for indigent families (World Bank, 2008).

In the context of electricity reforms, lower electricity lifeline tariffs were kept fixed while increases were concentrated on higher-consuming households in Armenia, Brazil, Kenya, and Uganda. In Kenya, a lifeline tariff was established for households that consume less than 50 kWh per month, a threshold commonly used in Africa as a subsistence-level benchmark. The lifeline tariff is estimated to be affordable to 99 percent of Kenyan households. Kenya also subsidized connection costs in place of electricity price subsidies, which helped expand coverage to poor households and those in remote and rural areas. Its rural electrification program to date has more than tripled the number of rural connections, and it has assisted households with connections by creating a revolving fund for deferred-connection fee payments and commercial bank loans for connection fees.

Affordable Alternative Energy

Providing an affordable alternative energy source can mitigate the impact of subsidy reform on low-income groups. A key objective of subsidies in many countries is to provide an affordable source of energy to low-income households. Subsidy reform can therefore often be more acceptable if it is accompanied by complementary measures that support this objective. Such measures were included in five reform episodes.

In Indonesia and Yemen, subsidy reform was facilitated by the government's efforts to help households convert from the use of kerosene for cooking to the use of low-cost LPG. In addition to being lower in cost, LPG produces lower levels pollution and CO_2 emissions. In Indonesia, LPG stoves and small LPG cylinders have been distributed free of charge.

Social Measures for SOEs Undergoing Restructuring

Subsidy reform involving SOE restructuring requires temporary sector-specific social measures to support employees and enterprises. In the short term, SOE restructuring may involve laying off part of the workforce or require increased investment in energy-saving technologies. Policies that mitigate the impact on workers and promote restructuring can increase support for subsidy reform. In the case of coal sector reform in Poland, unemployed miners had access to social assistance and job training.

In the context of fuel subsidy reform, the Iranian government undertook extensive consultation with enterprises to understand the challenges they faced if energy prices increased substantially. This led to a program targeted to agriculture and energy-intensive sectors hard hit by price increases, which included direct assistance and access to subsidized fuel. Such measures should be temporary, with a clearly specified life span, and should be communicated to the public to demonstrate the government's commitment to reforms.

Energy Pricing Should Be Depoliticized

Successful and durable reforms require a depoliticized mechanism for setting energy prices. Many countries have successfully implemented reforms only to see subsidies reappear when international oil prices increase. Out of 28 reform episodes, 11 were classified as partially successful because subsidies later reemerged. In Ghana, the 2005 reform eliminated fuel subsidies, but when oil prices soared in 2007 and 2008, the government abandoned its policy of linking domestic to international prices and automatic adjustment was temporarily suspended. In Indonesia, in spite of increasing international prices, subsidy reform reduced fuel subsidies from 3.5 percent of GDP in 2005 to 2 percent of GDP in 2006. However, unwillingness to fully pass through continued increases in international prices resulted in fuel subsidies escalating again to 2.8 percent of GDP in 2008.

Automatic Pricing Mechanisms

Automatic pricing mechanisms can help reduce the chances of reform reversal. Establishing an automatic pricing formula for fuel products can help distance the government from energy pricing and make it clearer that domestic price changes reflect changes in international prices that are outside the control of the government. Reliance on a formula can reassure the public that price increases would not lead to windfall profits for suppliers. South Africa has successfully implemented an automatic pricing mechanism for fuel products for over five decades. The mechanism's primary purpose there was to encourage private sector participation in the energy sector and secure an adequate supply of petroleum products despite the impact of sanctions on fuel supply during the apartheid era. Providing prices at least equal to import parity proved to be critical in incentivizing international firms to invest and maintain their activities in South Africa even during the anti-apartheid embargo.

The Philippines and Turkey successfully implemented such a mechanism during their transition to liberalized fuel pricing. Turkey launched its automatic pricing mechanism in 1998, setting a ceiling on the prices of almost all oil products based on international petroleum prices and the exchange rate. Initially, refineries and importers could set their own prices within this formula, although distribution companies and retailers could not. A full liberalization of fuel prices came into effect in 2005. In all three countries—South Africa, the Philippines, and Turkey—detailed information on the pricing mechanism and its implementation was disseminated to the public on government Web sites and through other media.

The adoption of such a mechanism is not a panacea for achieving a sustained reform of energy subsidies. A number of countries have abandoned these mechanisms shortly after adopting them, partly because of an unwillingness to pass sharp international price increases through to consumers. Gabon suspended its mechanism in August 2002 as international oil prices started to increase. Ghana adopted an automatic mechanism in February 2001 but suspended it before the end of the year. It reintroduced the mechanism in January 2003, only to suspend it again in June 2003. More recently, newly adopted pricing mechanisms have been suspended in other sub-Saharan African countries, including The Gambia, Sierra Leone, and Togo. The sustainability of these mechanisms can be enhanced if they are packaged and communicated as part of broader structural reforms, including expansion of targeted social safety net and social spending programs. Using price smoothing rules can also help to avoid large price increases.

Importance of Independent Bodies

Responsibility for implementing the automatic mechanism can be given to an independent body. Technical decisions on pricing can be delegated to an independent institution to ensure that subsidy reform proceeds as planned. The institution can also have the responsibility for implementing the automatic mechanism once subsidies are eliminated. A number of countries that successfully reformed subsidies for petroleum products (including South Africa and Turkey) and electricity (including Armenia, Kenya, the Philippines, and Turkey) gave responsibility for reforming and regulating energy prices to an independent agency.

In Turkey, the long process of market privatization in the petroleum industry had begun in 1990, but the full liberalization of prices was not achieved until 2005. Regulation of the petroleum product market was achieved with the passage in 2003 of the Petroleum Market Law, which transferred regulatory authority from the government to the Energy Market Regulatory Authority, an independent agency that was already regulating the electricity and natural gas markets. In addition to helping institutionalize the market economy, the Petroleum Market Law put Turkey in compliance with European Union (EU) legislation and other international obligations.

Adoption of a Smoothing Rule

A smoothing rule can be incorporated into the automatic pricing mechanism to avoid sharp increases in domestic prices (Coady and others, 2012). Countries often abandon automatic pricing mechanisms when international prices increase sharply. In China, for example, a key barrier to the adoption of an automatic pricing mechanism has been concern about the political and social consequences of fully passing through such sharp price increases. Some countries, including Chile, Colombia, Malawi, Nigeria, Peru, Thailand, and Vietnam, have used smoothing rules to address this problem.

Smoothing mechanisms can also help contain inflationary expectations if supported by appropriate macroeconomic policies. They can help dampen the effects of international price and exchange rate volatility. Several sub-Saharan countries, including The Gambia, Sierra Leone, and Togo, are considering the use of smoothing rules. With a smoothing mechanism, periods of sharp increases in international prices would only gradually be transmitted to domestic prices. For instance, energy price changes could be limited to a maximum of, say, 5 percent of the current consumer price in any given month.

To protect the budget over the medium term, smoothing must be applied both to price increases (when subsidies increase or taxes fall) and to price decreases (when subsidies decrease or taxes increase). How much smoothing the government chooses to implement will depend on its preference between lower price and higher fiscal volatility. Peru adopted a smoothing rule in 2004 whereby international price changes were fully passed through to domestic prices as long as the latter fell within a fixed price band. When prices fell outside this price band, the cost (if above) or benefit (if below) was absorbed by the budget. Since 2010, the band price limits have been updated to reflect trends in international prices, with adjustments limited to 5 percent. Although stabilization funds have also been used to smooth price increases, experience with such funds has been mixed, with funds exhausting their reserves during periods of sharp increases in international prices or incurring large contingent liabilities for the budget (Chile, Namibia, Peru, the Philippines, Thailand).

Role of the Government in a Liberalized Regime

Over the longer term, subsidy reforms for petroleum products should aim to fully liberalize pricing. More liberalized regimes—where prices are determined by private sector suppliers and move freely with international prices—tend to be more robust to the reintroduction of subsidies than automatic pricing mechanisms (Baig and others, 2007). Under a liberalized regime, the role of the government is to ensure that fuel markets are competitive and that there is free entry and exit from the sector. A well-functioning social safety net should be in place before countries liberalize to ensure that low-income groups can be protected from future price increases and thus avoid public pressure to reintroduce subsidies. Successful implementation of an automatic pricing mechanism can facilitate the transition to a liberalized pricing regime by getting the public used to frequent changes in domestic energy prices. It can also build up private suppliers' confidence that the government will not return to subsidized pricing. This approach was used in the Philippines, which adopted an automatic pricing mechanism in 1996 as part of its transition to a liberalized supply and pricing regime in 1998.

Continuing Role for Price Regulation in the Power Sector in Small Countries

In the electricity sector, the small size of the market in some countries limits the scope for competition and price liberalization. In many emerging and low-income economies, the electricity market is small. Under these circumstances, the market may not support many firms of a size sufficient to reap economies of scale and produce at the lowest possible cost. In such cases, price regulation will be needed, and competition alone will not be the best approach to reforming the sector (Besant-Jones, 2006). Prices should be determined by an autonomous agency and set at a level that is sufficient to avoid subsidies and ensure an adequate return to investment under efficient operations. Enhancing the progressivity of tariff structures by imposing higher tariff rates for larger consumers can also reduce subsidy expenditures while protecting the poor. For instance, there is scope to make tariff structures more progressive in many African countries. Greater emphasis could also be given to subsidizing connections rather than the consumption of electricity.

CHAPTER 5

Case Studies from the Sub-Saharan Africa Region

ANTONIO DAVID, FARAYI GWENHAMO, MUMTAZ HUSSAIN, CLARA MIRA, ANTON OP DE BEKE, VIMAL THAKOOR, AND GENEVIEVE VERDIER

PETROLEUM PRODUCT SUBSIDIES

Ghana

Context

Ghana is a country of over 24 million people, rich in natural resources, including arable land and minerals. The country recently discovered offshore oil reserves, and 2011 was the first full year of production. Although Ghana's oil reserves are relatively small on a global scale—with production from the current Jubilee field expected to peak at 120,000 barrels a day—there is considerable upside potential from new discoveries. Moreover, Ghana is in the process of building up infrastructure for the commercial use of its gas reserves with potentially significant benefits in reducing energy costs and developing downstream industries.

Since 2004, deregulation has allowed oil marketing companies to enter the market for importing and distributing crude oil and petroleum products. Until that time, the Tema Oil Refinery (TOR) had a monopoly on the production and importing of refined products. Since then, deregulation has allowed oil marketing companies to enter the market for importing and distribution of crude oil and petroleum products. Under the current system, a pricing formula exists for all petroleum products. The current price-adjustment mechanism is the result of 2005 reforms, although it has not always worked as originally envisaged. The National Petroleum Agency (NPA), also established in 2005, reviews fuel prices twice a month. It provides recommendations to the Minister of Energy on adjustments to cost-recovery levels, based on a backward-looking formula incorporating changes in world fuel prices in the preceding two weeks.

The decision to adjust pump prices is at the discretion of the executive. If price increases are warranted but not implemented, the cost of subsidies is in principle borne by the budget. However, in the past, TOR carried the cost of the subsidy, and underpricing of petroleum products saddled TOR with large losses that spilled over into the financial sector in the form of nonperforming loans. The government was forced ultimately to clear TOR's arrears to the banking sector at a large budgetary cost. Since October 2010, a hedging scheme using call options has also

TABLE 5.1

Ghana: Key Macroeconomic Indicators, 2000–2011

	2000	2003	2008	2010	2011
GDP per capita (US$)	400	563	1,266	1,358	1,580
Real GDP growth (percent)	4.2	5.1	8.4	8.0	14.4
Inflation (percent)	25.2	26.7	16.5	10.7	8.7
Overall fiscal balance, cash (percent of GDP)	−6.7	−3.3	−8.5	−7.2	−4.1
Public debt (in percent of GDP)	123.3	82.8	33.6	46.3	43.4
Current account balance (percent of GDP)	−6.6	0.1	−11.9	−8.4	−9.2
Oil imports (percent of GDP)	−7.1	−5.0	−8.3	−6.9	−8.3
Oil exports (percent of GDP)	0.0	0.0	0.0	0.0	7.2
Oil consumption per capita (liters)	n.a.	91.1	91.4	98.7	110.7
Poverty headcount ratio at US$1.25 per day (PPP) (percent of population)	39	n.a.	30	n.a.	n.a.

Sources: International Energy Agency (IEA); IMF, *World Economic Outlook* (WEO); World Bank, *World Development Indicators*.
Note: n.a. = not applicable; PPP = purchasing power parity.

provided some temporary protection against upward movements in oil prices. The government purchases monthly call options that generate revenues in the event of upside shocks to global oil prices; these revenues are used to cover temporary delays in adjusting domestic petroleum product prices to cost-recovery levels (IMF, 2011a).

Reforms Since 2001

The past decade has been marked by several attempts to deregulate fuel prices in Ghana (Figure 5.1). In 2001, a 91 percent adjustment of petroleum pump prices was driven in part by the desire to restore TOR's financial health. Delays in adjusting petroleum prices during 2000 led to large accumulated losses for the state-owned public energy company, which reached 7 percent of GDP (IMF, 2001). The reform was soon abandoned, however, in the face of rising world prices and a depreciating currency. TOR's losses were largely absorbed by the state-owned Ghana Commercial Bank, whose solvency was threatened.

In early 2003, recognizing the unsustainable financial position of both TOR and Ghana Commercial Bank, the government renewed its commitment to cost-recovery pricing with a 90 percent increase in pump prices. Facing widespread opposition to the price increase, the government partially reversed the price increase in the run-up to the 2004 elections, and it abandoned cost-recovery adjustments until 2005. In 2004, the subsidies to TOR reached 2.2 percent of GDP, and the company continued to borrow from Ghana Commercial Bank to finance its operations (IMF, 2005a).

Strategic and Mitigating Measures

The deregulation of petroleum product pricing in 2005 was accompanied by strategic measures meant to ensure broad popular support for the reform. The strategy was supported by research, communication, and programs to mitigate the impact on the most vulnerable groups, all of which contributed to its successful implementation.

- *Research.* A poverty and social impact assessment studying the impact of fuel subsidy removal revealed that the program was poorly targeted, with the rich receiving the lion's share of the benefits (Coady and Newhouse, 2006).

- *Communication.* The government engaged in a widespread communications campaign, including public addresses by the president and the Minister of Finance, explaining the reform's benefits. The results of the poverty and social impact assessment were made public and discussed in a dialogue with various stakeholders, including trade unions. The government also explained how resources freed from subsidizing energy products would partly be reallocated to social priorities (Global Subsidies Initiative, 2010).

- *Assistance to the poor.* The government introduced a number of programs aimed at mitigating the effect on the most vulnerable, including the elimination of fees for state-run primary and secondary schools; an increase in public-transport buses; a price ceiling on public-transport fares; more funding for health care in poor areas; an increase in the minimum wage; and investment in electrification in rural areas.

The administration of the publicly released price-adjustment formula was transferred to the newly established NPA. The delegation of regulatory powers to the NPA was meant to isolate the decision to adjust prices from political intervention. Prices were adjusted by an average of 50 percent, and the government remained committed to regular adjustment for several years. In the wake of the

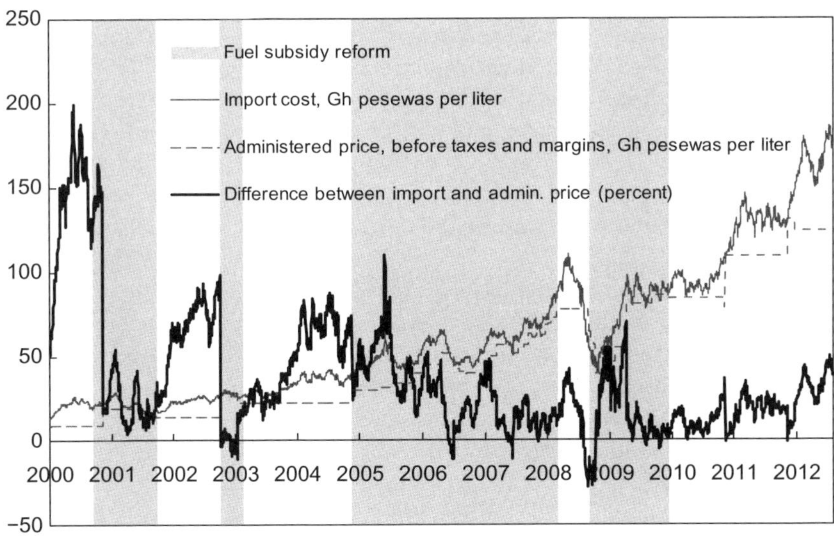

Sources: IMF staff estimates; National Petroleum Agency (Ghana).
Note: admin. = administered.

Figure 5.1 Ghana: Fuel Price Developments, 2000–2012
The 2005 reforms ushered in a period of market-based fuel pricing. However, political considerations have at times interfered with this process.

2007–8 global fuel and food crisis and in the run-up to the 2008 elections, however, automatic adjustment was temporarily suspended.

The NPA remains the main regulatory agency and publishes the price adjustments required for cost recovery on a biweekly basis. When an upward price adjustment has been required in recent years, the shortfall has often been covered by the budget or, more recently, by hedging profits. This has resulted in infrequent and large price adjustments when hedging profits were exhausted and the fiscal burden became too onerous. Prices were adjusted twice in 2011, by 30 percent in January and by 15 percent in December. Prices were not adjusted in 2012, with the exception of a small downward adjustment early in the year, and the gap between domestic and global oil prices, exacerbated by a depreciating currency, has increased substantially (IMF, 2012b and 2012c).

Lessons

A number of lessons can be drawn from Ghana's experience in the past decade.

The durability of reform depends crucially on political will and the independence of regulatory agencies from political interference. Without these conditions, it is difficult to maintain an independent regulatory agency. The NPA is not free to adjust prices without the consent of the executive: it has adjusted prices only three times (once downward) since January 2011. Although democratically elected governments have stronger mandates to implement difficult reforms, commitment to automatic adjustment often falters in the run-up to elections.

A constant dialogue with stakeholders and civil society at large about the cost of subsidies is necessary to maintain commitment to the reform. Recent attempts at adjusting prices have not been accompanied by an extensive public information campaign similar to the 2005 efforts. Price increases have been irregular, difficult to anticipate, and usually announced shortly before being implemented. This can result in strong opposition by various stakeholders, including powerful trade unions, and can undermine the government's efforts. The 2005 campaign was also successful because it engaged civil society and powerfully demonstrated the cost of fuel subsidies by sharing the results of the poverty and social impact assessment.

Supportive research and analysis are important for convincing the public of the benefits of reforms. During the 2005 reform, the poverty and social impact assessment was crucial in demonstrating the costs of subsidies. The assessment also outlined that fuel subsidies were a poor policy measure in the fight against poverty: in Ghana, less than 2.3 percent of outlays on fuel subsidies benefited the poor.

Visible mitigating measures increase the likelihood of success. Although fuel subsidies are ill targeted, they are a direct transfer to most if not all citizens, their benefits are immediate and easy to understand compared with other social programs, and the individual cost of their removal is swift and substantial—particularly for the poor who have no income cushion, unless they receive alternative compensation. A key element of successful reform, therefore, is the efficient and visible reallocation of the resources saved through the removal of fuel subsidies to programs with immediate benefits to the most vulnerable. In Ghana, an expansion of cash transfers through the Livelihood Empowerment against Poverty (LEAP) program and additional spending on health and education subsidies would be good candidates.[1]

Namibia

Context

Namibia is one of sub-Saharan Africa's richest countries and has a relatively stable macroeconomic environment. Income inequality and unemployment are very high, however. Mineral exports, transfers from the Southern African Customs Union, and prudent fiscal policy in the past have helped the Namibian government to sustain economic growth while maintaining fiscal and current account surpluses. Inflation in Namibia is closely linked to South Africa's inflation (its currency is pegged to the South African rand) and has remained within single digits since reaching a peak of 11.9 percent in August 2008, driven by a surge in international oil prices. The Namibian economy is sensitive to changes in international fuel prices because of the relative importance of energy-intensive industries, such as fishing and mining.

Namibia is characterized by political stability and a relatively well-functioning democracy. The ruling political party is dominant and has won elections with large majorities since independence in 1990. Labor unionization is fairly high and the largest trade union federation, the National Union of Namibian Workers, is a strong political ally of the ruling party.

Namibia has a wide range of formal publicly funded social welfare programs. Social security, welfare, and housing spending averaged 5 percent of GDP during

TABLE 5.2

Namibia: Key Macroeconomic Indicators, 2000–2011

	2000	2003	2008	2010	2011
GDP per capita (US$)	2139.7	2607.9	4276.0	5244.1	5828.2
GDP growth (percent)	4.1	4.3	3.4	6.6	4.9
Inflation (percent)	9.3	7.2	10.4	4.5	5.8
Overall fiscal balance (percent of GDP)*	−0.9	−6.1	2.4	−4.2	−11.3
Public debt (percent of GDP)*	20.4	26.4	18.2	16.2	27.4
Current account balance (percent of GDP)	7.9	6.1	2.8	0.3	−1.7
Oil Imports (percent of GDP)	3.5	4.5	2.4	5.3	5.9
Oil exports (percent of GDP)	0.0	0.0	0.0	0.0	0.0
Oil consumption per capita (liters)	n.a.	491.5	596.2	731.0	812.9
Poverty headcount ratio at US$1.25 per day (PPP) (percent of population)	n.a.	31.9	n.a.	n.a.	n.a.

Sources: IEA; IMF, WEO; World Bank, *World Development Indicators*.
*Figures are for the fiscal year, which begins April 1.

---

[1]According to the World Bank (2012a), LEAP is among the most well targeted of programs for the poor. Fuel subsidies, in contrast, which disproportionately benefit those in higher income brackets, reached weekly levels in May and August 2012 that matched the annual budget contribution for LEAP.

2005–11. The government's income support grants include a universal social pension system for the elderly and the disabled, a variety of grants for children, labor-based work programs, and shelter and housing programs. Despite some weaknesses, such as errors of inclusion and exclusion, anecdotal evidence suggests that Namibia has a well-targeted social safety system.

The downstream market for liquid fuels in Namibia is administered through acts of parliament that set out clear parameters to calculate fuel prices. According to the acts, the prices of petrol and diesel are regulated, whereas the prices of all other petroleum products are determined by market forces. The country has no refining capacity and imports its refined fuels mainly from South Africa through the port of Walvis Bay. The Ministry of Mines and Energy regulates the industry while the Namibian Petroleum Corporation (Namcor), a state-owned enterprise, acts as an operational arm of the government in the market. There are five private companies involved in the marketing of petroleum products: BP, Caltex Oil, Engen, Shell, and Total. Each private company supplies its own network of distribution outlets, but all share import and storage facilities at Walvis Bay. In 1999, Namcor was mandated by the government to import 50 percent of Namibia's petroleum, leaving the other 50 percent for private companies. That share was recently reduced as a result of Namcor's operational difficulties.

Price setting of fuel pump prices for diesel and petrol is based on a formula with three components: basic fuel price, based on the international spot price; domestic fuel levies and taxes; and the so-called slate account, which is essentially used to smooth volatility in local pump prices. The slate account, monitored by the Ministry of Mines and Energy, is a notional record used to keep track of the degree of under- or overrecovery by fuel-importing private companies. However, the price formula is not completely automatic, as the ministry has some discretion on how much pass-through to allow with underrecoveries absorbed by the slate account.

Fuel Price Reforms in the 1990s and 2000s

According to the Ministry of Mines and Energy, the original motivations for deregulating fuel prices in Namibia were to eliminate fuel subsidies, paid out of the National Energy Fund (NEF), and to respond more efficiently to changes in international oil prices. Several problems associated with the managed petroleum and petrol-product scheme may have motivated the reforms (Amavilah, 1999). First, the NEF compensation scheme came with fiscal costs amounting to about N$170 million between 1990 and 1996, about 0.2 percent of GDP (Figure 5.2). Although the fiscal costs paid out of the NEF seem small in percent of GDP, they do not include transfers that may have been paid directly to Namcor or quasi-fiscal costs arising from losses incurred by the company. Namcor sometimes receives direct transfers from the government because it does not participate in the slate program and is therefore not compensated for underrecovery through the slate account. The subsidies may also have reduced incentives for petroleum firms to improve their efficiency to help offset their losses.

After the adoption of the new price mechanism, the slate account is supposed to be balanced through price adjustments, in theory. In particular, the price adjustment formula should adjust prices so that the value of the cumulative slate balances is kept within a predetermined level of N$3 million. In practice, however, balancing the slate account has sometimes involved transfers from the budget to the NEF and then to the slate account (see Figure 5.2). The wholesale prices of all petrol grades and diesel are published in a government gazette at each price adjustment. Tax revenue data is published in budget documents.

The Ministry of Mines and Energy used a structured, balanced, and consultative approach to price deregulation and subsidy removal. The National Energy Council, chaired by the Minister of Mines and Energy, established the National Deregulation Task Force in 1996 to examine fuel price deregulation through a consultative process. This culminated in the publication of a white paper on energy policy in 1998, which articulated, among other issues, the importance of keeping targeted subsidies to remote areas, gradual deregulation, and enhancing transparency in government fuel tax revenues. The fuel price mechanism with quarterly price reviews was adopted in 1997.

NEF expenditures to cover subsidies started to decline only after 2001. That represented a full three years after the release of the white paper, an indication that the implementation of fuel subsidy removal takes time. In addition, as shown

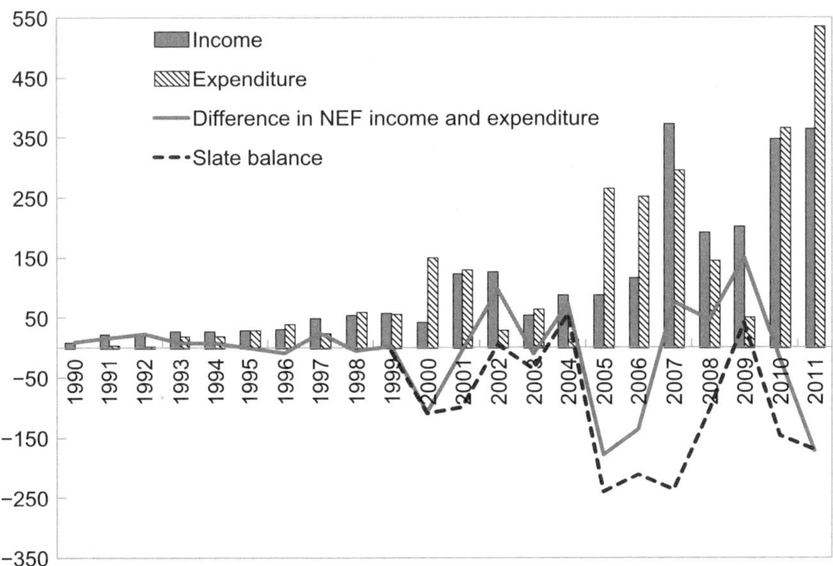

Source: Bank of Namibia, *Quarterly Bulletin*, March 2005.
Note: NEF = National Energy Fund.

Figure 5.2 Namibia: National Energy Fund and Slate Account, 1990–2011 (Millions of Namibian dollars)
Resources devoted to smoothing out fuel prices experienced sharp swings over time.

by the slate balance in Figure 5.2, close to full cost recovery by private firms came only after 2001.

Domestic fuel prices in Namibia increased steadily from 2003 onward and more than doubled from early 2007 to a peak in July 2008. In response to the 2007–8 fuel price shocks, the authorities replaced the quarterly fuel price adjustments with monthly fuel price reviews so as to increase pass-through. However, the Ministry of Mines and Energy did not allow retail prices to rise as fast as world prices, transferring funds from the NEF to the private petroleum firms to compensate them for keeping prices below cost-recovery and thus subsidizing users, including the powerful interest group of taxi drivers. However, in July 2008, the ministry announced that the NEF had come under financial pressure as a result of underrecoveries and was no longer in a position to cushion increasing fuel prices.

Overall, although fuel prices have generally moved in line with international oil prices, the government has from time to time accommodated pressures to limit the full pass-through of changes in international prices. In the 2006–7 budget, the government made a one-off budgetary provision of N$206 million (0.4 percent of GDP) to offset the NEF's accumulated losses. The government also faces contingent liabilities arising from Namcor's operational losses. In 2009, Namcor had operational losses of N$257 million, prompting the government to award it a N$100 million grant and a bailout package to the tune of N$260 million (0.5 percent of GDP) as well as a portion (N$0.08 per liter) of the existing fuel levy to help boost the state-owned oil corporation's finances. More recently, Namcor lost its mandate to supply 50 percent of Namibia's total fuel requirements in February 2011 because of operational difficulties.

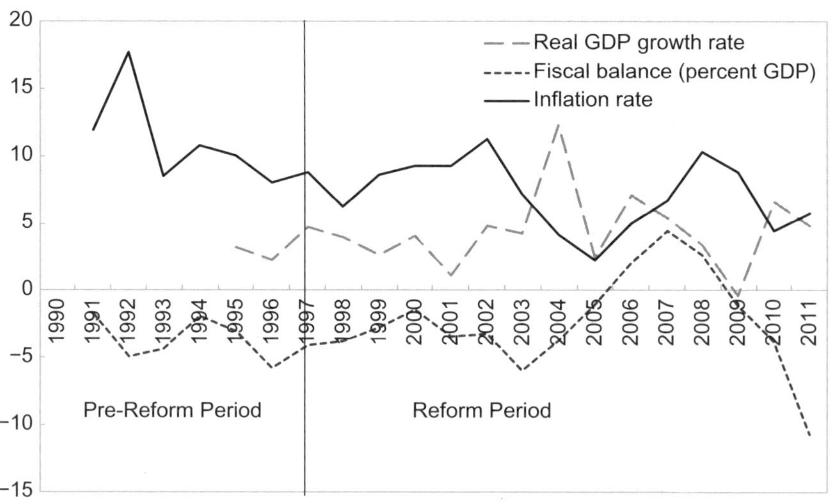

Source: Namibian authorities.

Figure 5.3 Namibia: Macroeconomic Developments and Fuel Subsidy Reform, 1990–2011
The fuel subsidy reform helped to consolidate Namibia's macroeconomic stability.

Mitigating Measures

The fuel price smoothing mechanism has been complemented by several mitigating measures to address the increases in fuel prices. Unlike its Southern African Customs Union (SACU) counterparts, Namibia did not experience violent protests in response to rising fuel and food prices, although taxi drivers complained when fuel prices increased. This might be partly explained by the Ministry of Mines and Energy's fuel price smoothing mechanism and other mitigating measures that were put in place in 2008 to address poverty and alleviate the temporary impact of high fuel and food prices. Mitigating measures included a zero-rate value-added tax on selected food items, rebate facilities for food importers, and a food distribution program to feed the most vulnerable. In addition, rural pump prices are subsidized as part of the socioeconomic policy of government. This is achieved by subsidizing transportation costs to remote areas to ensure that the pump price in remote areas is not inflated by retailers' transport costs. Claims on actual road deliveries are submitted by the oil companies to the Ministry of Mines and Energy for reimbursement from the NEF.

Lessons

Comprehensive planning and gradual implementation were key to success. The Namibian authorities undertook comprehensive planning, which included broad consultation with civil society, culminating in a comprehensive reform plan that retained a targeted subsidy for remote areas.

Reforms were implemented gradually, allowing enough time for consensus building between the government and various stakeholders.

Price adjustments that employed smoothing mechanisms helped prevent social unrest. The reform established a quarterly (later monthly) price adjustment mechanism in line with changes in international prices but incorporating a price smoothing mechanism to avoid sharp price adjustments. This, along with the introduction of other mitigating measures, allowed Namibia to manage the large price shocks of 2008 and 2011 with no social unrest.

Depoliticization of the price adjustment mechanism has been made difficult by legal obligations to the state-owned energy company. The legally stipulated participation of the state petroleum company in the importation and supply of petroleum products seems to have prevented a full depoliticization of the price adjustment mechanism (i.e., allowing prolonged underrecoveries). This in turn has resulted in large losses for the company that have had to be covered by fiscal transfers. This suggests the need to carefully design price smoothing mechanisms.

Niger

Context

Niger is a large and landlocked country that is extremely vulnerable to external shocks, mostly to climatic conditions and commodity prices. In the past decade, growth has been slowly gathering momentum, though it has also suffered important setbacks. Niger's medium-term growth potential is linked to the expansion

TABLE 5.3

Niger: Key Macroeconomic Indicators, 2000–2011

	2000	2003	2008	2010	2011
GDP per capita (US$)	155.0	223.8	361.0	363.6	420.7
GDP growth (percent)	−2.6	7.1	9.6	10.7	2.2
Inflation (percent)	2.9	−1.8	10.5	0.9	2.9
Overall fiscal balance (percent of GDP)	−3.8	−2.8	1.5	−2.4	−3.0
Public debt (percent of GDP)	118.8	90.1	21.0	23.7	29.2
Current account balance (percent of GDP)	−6.7	−7.5	−13.0	−19.9	24.7
Oil imports (percent of GDP)	4.0	2.4	3.8	4.7	4.7
Oil exports (percent of GDP)	0.0	0.0	0.0	0.0	0.0
Oil consumption per capita (liters)	n.a.	n.a.	36.4	33.1	34.3
Poverty headcount ratio at US$1.25 per day (PPP) (percent of population)	n.a.	n.a.	43.6	n.a.	n.a.

Sources: IEA; IMF, WEO; World Bank, *World Development Indicators.*

occurring in the oil and mining (uranium) sectors. The country recently became a fuel exporter, and uranium production is expected to double in the near future with the coming onstream of an important mine currently under development. In addition, the country has the potential to become a crude oil exporter, with five new oil production sharing agreements just signed. A new pipeline to link Niger with the Chad–Cameroon pipeline is envisaged.

Niger ranks at the bottom of the United Nations Development Program's Human Development Index, with per capita GDP in purchasing power parity (PPP) terms of US$720 in 2010, one of the lowest in the world. Niger's government is highly centralized. The current authorities have been in power since April 2011, following a one-year transition to democracy after a February 2010 coup d'état. Since then, the political situation has been stable, although according to the World Bank (2012b, page 2), there is a risk of political fragility "where failure of the government to deliver tangible results could result quickly in the loss of popular support and a political stalemate."

With the start of operations of its new oil refinery, SORAZ, fuel imports have come nearly to a halt since early 2012. Niger was an oil importer until the end of 2011. Its market size is small, with annual domestic consumption of about 7,000 barrels a day. The state-owned company SONIDEP has a monopoly on imports and distribution. The new refinery is expected to reach a maximum capacity of 20,000 barrels per day of fuel, including gasoline, diesel, and liquefied petroleum gas (LPG). About one-third of the petroleum products produced by SORAZ feeds the domestic market, with the remainder being exported. SONIDEP is in charge of marketing the petroleum products.

This case study focuses on the period in which Niger was an oil importer, until the end of 2011. It builds on IMF technical assistance support provided to Niger in 2001 to elaborate a pricing formula akin to a full pass-through rule for the automatic adjustment of the price of imported petroleum products. In 2010, a note was prepared by the IMF Fiscal Affairs Department to support the authorities in their intention to eliminate the posttax fuel subsidies, in the context of discussions with the IMF to prepare an assessment letter.

Fuel Price Reforms Since 2001

According to the formula established with the help of technical assistance from the IMF in 2001, automatic pass-through of international prices would be achieved through a flexible, transparent, and automatic mechanism. The retail price would be adjusted monthly whenever the change in international prices was above CFAF 5. Otherwise, the price at the pump would not change, and taxes would counteract the increase or decrease in prices. The pricing formula included fuel import costs (c.i.f. import price at the port); estimated costs and margins of importing and distributing fuel to domestic consumers (storage and distribution margins); and net fuel taxes (ad valorem customs and value-added taxes and specific excise taxes). A multisectoral body was envisaged to be statutorily in charge of applying the formula; however, such a body was never created.

As international prices started to increase in 2005, an explicit subsidy component was introduced in the formula. The subsidy was initially used to smooth domestic prices. Then, as international import prices increased rapidly and steadily up to mid-2008, the subsidy component rose in order to keep domestic retail prices fixed for extended periods. The increase in international prices and the depreciation of the euro resulted in a significant increase in the subsidies in 2010. Because fuel prices were substantially lower in Niger than in some neighboring countries, increased smuggling contributed to a strong rise in fuel imports.

Changes in import prices without corresponding pass-through to retail prices resulted in a reduction of government tax revenue from fuels. The net fiscal contribution of fuel taxes decreased from 1 percent of GDP in 2005 to 0.6 percent in 2009 and to 0.3 in 2010. The cost of the subsidy on petroleum products amounted to more than 1 percent of GDP. Although this pattern has applied to all products, the tax decline in the case of gasoline was more pronounced, going from a peak of 0.8 percent of GDP in 2005 to 0.3 percent of GDP in 2009. Net taxes on diesel also declined from 0.3 percent of GDP in 2005 to 0.2 percent of GDP in 2009. The net tax on kerosene has been continuously negative over this period, although the fiscal cost of this measure has been limited, because the share of kerosene consumption is fairly low.

When the subsidy reached unsustainable levels, the authorities decided to start implementing a strategy to gradually phase out subsidies. The size of the subsidy, together with its very regressive distributional impact, was a critical factor in the authorities' decision to eliminate it. Indeed, the population groups that benefited more from the subsidy were the higher-income groups, who consumed more gasoline. Although this is particularly the case in gasoline consumption, it is less so in kerosene and lamp oil, which are more widely consumed by lower-income groups. Fuel prices were increased by 12 percent in mid-2010 (Figures 5.4 and 5.5).[2]

---

[2] Weighted average of the prices of gasoline, kerosene, and diesel. Full pass-through includes import prices, taxes, and margins in the formula. In both cases, price increases were considered preconditions for the IMF to issue an assessment letter and to proceed with the Extended Credit Facility–supported program review.

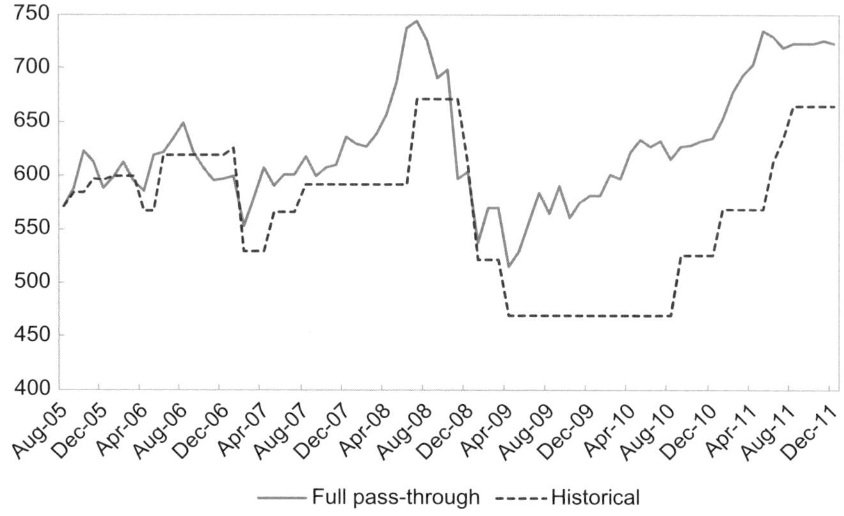

Sources: IMF Fiscal Affairs Department data; and authorities.

Figure 5.4 Niger: Fuel Price Developments, 2005–11 (Central African francs per liter)
Domestic fuel prices have tended to follow international prices with a lag.

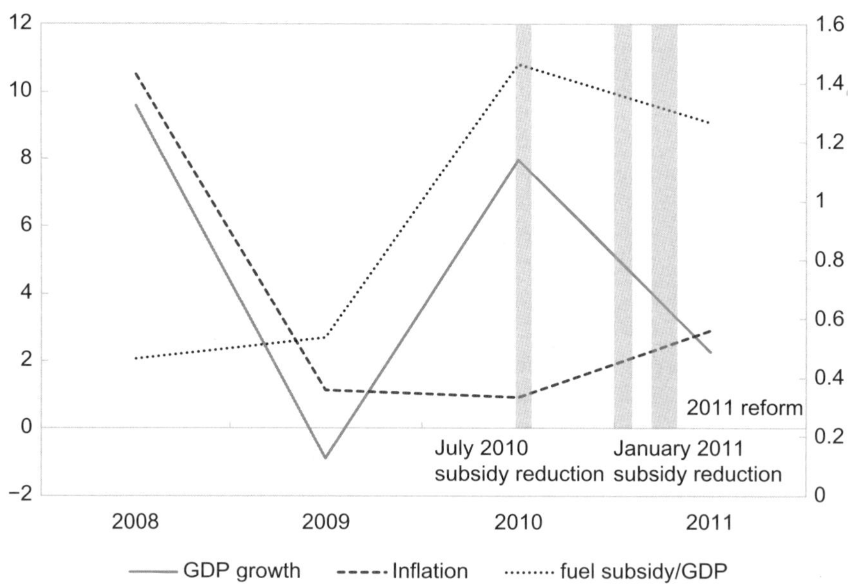

Source: IMF staff estimates.

Figure 5.5 Niger: Macroeconomic Developments and Energy Subsidy Reforms, 2008–11 (Percent of GDP or rate)
Niger has tried to rein in fuel subsidies in the context of volatile macroeconomic performance.

The agreed reform contained two steps. First, international oil price variations would be passed through to domestic prices starting in June 2011. Second, the existing subsidy would be gradually unwound over the following 12–18 months. Fuel prices were increased by about 8 percent in mid-2011. As a result, the subsidy was significantly reduced, though not completely eliminated, and the total amount devoted to fuel subsidies in 2011 was kept below the 2010 level (1.1 percent of GDP).

Country-specific circumstances and the political situation played key roles in the design and pace of the reform. First, the imminent start of domestic fuel production introduced urgency in the phasing out of the subsidies. The authorities thought that it would have been politically unacceptable to increase prices exactly when domestic production was starting. In fact, the society was expecting rather the opposite: a decrease in fuel prices with the start of domestic production. Second, the initial reforms (in late 2010 and early 2011) were implemented by a transitional government that believed it had less legitimacy to embark on such a sensitive reform process.

To increase public awareness about the dimension of the problem, for the first time the budget explicitly reflected the costs of the subsidy. This helped create an appropriate environment for the subsidy's elimination. In addition, and to help overcome vested interests and gain support from the civil society, the government introduced public information campaigns pointing out the regressive nature of the subsidies and linking the savings from petroleum price increases to priority social spending.

The authorities opted for a consensual approach to the reform, incorporating all relevant shareholders. They established a committee (the Comité du Différé) to discuss the best way to approach the reforms and their subsequent implementation. In this context, dialogue and consensus building were key to the positive outcome of the process.

As a result of the reform, retail prices started increasing in June 2011 and continued increasing through August 2011, but they remained fixed again from September until the end of the year. Indeed, the monthly cost of the subsidy reached nearly CFAF 4 billion in May 2011, to be reduced to half from August onward. The authorities decided to stop the price increases in September because they believed the prices were then aligned with prices within the region.

However, prices were set below international prices once Niger started producing fuel domestically. As a result of an agreement between the authorities and the foreign investor in the petroleum sector, SORAZ started selling its fuel products at CFAF 336 per liter for gasoline and CFAF 340 for diesel, which were below the international prices. The prices were fixed for the first six months of operation of the refinery, with refined products' prices supposed to be set by a formula linked to world market prices after that period. Nonetheless, the prices did not change. More recently, an agreement has been reached between the government and the transportation trade unions aimed at developing proposals to further lower retail fuel prices. As a result, the fuel tax (*taxe intérieure sur les produits pétroliers*, or TIPP) will be reduced from 15 to 12 percent starting in 2013.

The overlap of the subsidy reform with the start of fuel and oil production makes Niger a very special case. As a result, it is difficult to assess at this stage how durable the fuel subsidy reform would have been if domestic production had not started at the same time.

Mitigating Measures

The more recent fuel price reform was accompanied by mitigating measures to protect the poorest segments of the population from increases in transportation costs:

- *Subsidy to the transport sector.* Following negotiations with the civil society and private sector operators, a direct subsidy to the transport sector was introduced (*tickets modérateurs*), because this sector was the most affected by the increase and the poorer sectors of the population were the ones that used more public transport. The costs of the subsidy policy were still reduced significantly because the costs of the mitigating measures (less than 0.1 percent of GDP) were significantly lower than the subsidy itself.

- *Increased social spending, emphasizing education.* The discontinuation of the subsidy on fuel products created room for a 19 percent increase in social spending in the 2012 budget compared with 2011, with particular emphasis on investment in education. The public wage bill was increased to accommodate the recruitment of 4,000 teachers in early 2012.

Lessons

There is a need to appropriately understand the extent of the fuel subsidy problem. Determining the distributional incidence of the subsidies can also help to ensure commitment to the reform.

Promoting an understanding of the issues by society as a whole is important. Being transparent about the costs of the subsidy by using an explicit budget line proved very useful in Niger. An adequate public information campaign also played a crucial role in ensuring the support of the society for reform. In Niger, there were debates on television and radio about this issue.

A participative approach is valuable. Adopting a participative approach to decision making was also useful, particularly through the establishment of an ad hoc and inclusive committee.

Sufficient time needs to be allowed to build support. There is a need for sufficient time to explain, negotiate, and implement the reform. Building reform momentum, stakeholders' consensus, and social support requires time. In the case of Niger, ensuring that all stakeholders were on board and agreed with the main elements of the reform took about six months.

Engaging financial partners can be helpful. Engaging partners can help to ensure that there is sufficient information about the problem and create pressure to launch the reform process. A delicate equilibrium needs to be reached between encouragement and ownership of the reform process.

Ensuring that mitigating measures reach the most affected groups is crucial. These measures can take the form of targeted subsidies based on a detailed analysis of who would be the most affected vulnerable groups.

Fuel subsidy reform becomes more complicated when a country becomes an oil exporter. At such times, it might be more difficult to resist the expectations and pressures from civil society to significantly lower pump prices.

Nigeria

Context

Nigeria is the world's fifth leading oil-exporting country. The oil and gas sector accounts for around 25 percent of GDP, 75 percent of general government fiscal revenues, and over 95 percent of total exports. Nigeria's federalist fiscal relations are quite complex and driven by substantial (and constitutionally mandated) oil revenue sharing among the federal government, 36 (oil-producing and non-oil-producing) states, and various local governments.

Nigeria has administratively set maximum prices for kerosene and gasoline and an indicative price for diesel.[3] At the core of this system, which was established in 2003, is the Petroleum Products Pricing Regulatory Agency, which sets these prices every month. This agency applies import parity but is also expected to stabilize prices, which it does with the help of the Petroleum Support Fund (PSF). When total costs are below the maximum price, the marketer benefits from what is called an "overrecovery"; if they are above, there is an "underrecovery." Any overrecoveries are to be paid into the PSF, supplementing the funds appropriated from the budget, whereas underrecoveries would be compensated from the PSF. The Petroleum Products Pricing Regulatory Agency posts Product Pricing Templates for kerosene and gasoline on its Web site. They show the

TABLE 5.4

Nigeria: Key Macroeconomic Indicators, 2003–11

	2000	2003	2008	2010	2011
GDP per capita (US$)	390.0	524.3	1401.2	1465.1	1521.7
GDP growth (percent)	5.3	10.3	6.0	8.0	7.4
Inflation (percent)	6.9	14.0	11.6	13.7	10.8
Overall fiscal balance (percent of GDP)	12.4	−4.3	1.7	−4.2	0.1
Public debt (percent of GDP)	84.2	63.9	11.6	15.5	17.2
Current account balance (percent of GDP)	12.5	−5.9	14.1	5.9	3.6
Oil imports (percent of GDP)	5.1	2.5	5.2	4.9	7.9
Oil exports (percent of GDP)	49.8	39.2	40.6	32.7	36.9
Fuel consumption per capita (liters)	n.a.	98.6	88.0	79.2	93.5
Poverty headcount ratio at US$1.25 per day (PPP) (percent of population)	n.a.	n.a.	n.a.	33.7	n.a.

Sources: IEA; IMF, WEO; Nigerian authorities; World Bank, *World Development Indicators*.

---

[3] Diesel was deregulated in 2007 and is not subsidized.

maximum prices but also the estimated costs of importing fuel—the so-called landing costs—as well as the costs of domestic distribution, decomposed into trading margins and fees, all of which are regulated.

Nigeria has subsidized kerosene and gasoline at a substantial cost to the government. Domestic fuel price setting has never been responsive enough to changing international prices. Importers have typically been unable to recover costs, and so from the beginning the PSF never received payments, only made them. As the gap between the administered price and the import parity price increased, subsidy costs rose from 1.3 percent of GDP in 2006 to 4.7 percent of GDP in 2011. In 2011, the budget appropriation for the PSF was just 0.6 percent of GDP, and funding for the subsidies came from Nigeria's oil stabilization fund (the Excess Crude Account). The price gap has encouraged widespread smuggling to neigh-

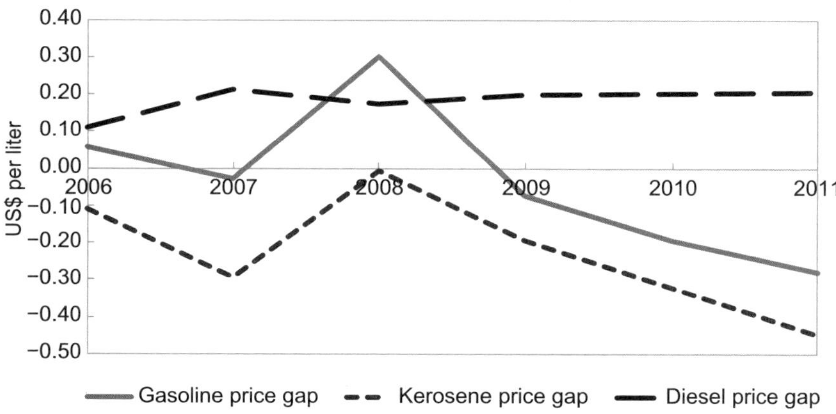

Sources: IMF staff calculations.

Figure 5.6 Nigeria: International and Domestic Fuel Prices, 2006–11 (Difference between world price and domestic price)
Domestic fuel prices in Nigeria have recorded substantial gaps relative to international prices.

TABLE 5.5

Nigeria: Developments in Fuel Prices and Fuel Subsidies, 2006–12

	2006	2007	2008	2009	2010	2011 Est.	2012 Proj.
Fuel subsidy (billion naira)*	251	290	637	399	797	1,761	1,570
Fuel subsidy (percent of GDP)*	1.3	1.4	2.6	1.3	2.3	4.7	3.6
Fuel prices (naira per liter)							
Diesel (deregulated)	81	90	118	94	112	152	144
Kerosene (subsidized)	50	50	50	50	50	50	50
Gasoline (subsidized)	65	70	70	65	65	65	97

Sources: IMF staff calculations and projections; Nigerian authorities.
* For 2012, includes one-off payment of about 1 percent of GDP to settle arrears accrued in 2011. Est. = estimate; Proj. = projection.

boring countries and other abuses (e.g., overinvoicing of gasoline imports), which have contributed to the escalating costs.

The subsidy regime has also been a disincentive to investment in domestic refining capacity. None of the 20 refinery licenses issued since 2000 have been used. Although Nigeria produces some 2.5 million barrels of oil per day, it is heavily dependent on the import of fuel products. Its four state-owned refineries, operating sometimes at only about 20 percent of capacity and rarely above 40 percent, meet only about 20 percent of the domestic demand.

Reform Since 2011

In mid-2011, the government decided to radically curtail gasoline subsidies and waged a public campaign the rest of the year to convince the population. The debate on removal of fuel subsidies was initially supported by several state governors, who wanted to free up resources to be able to pay their civil servants the new minimum wage. This proposal was hotly debated in the press, by business and civil society groups, and it was debated in the National Assembly during the rest of the year, with the government strongly trying to make a convincing case. On January 1, 2012, the price of gasoline was raised to a cost-recovery level—a 117 percent increase. The price of kerosene, a cooking fuel used mainly by poorer households, was not changed. However, in response to intense social unrest, the government scaled back the price increase to 49 percent by mid-January. Evidently, despite six months of debate, the measure did not enjoy sufficient public support.

The main plank in the government's campaign for the subsidy removal was the Subsidy Reinvestment and Empowerment (SURE) Program. The SURE program was announced only in November. It was preceded by public statements by the president and by budget documents (e.g., the 2012–15 Medium-Term Expenditure Framework and the Fiscal Strategy Paper) highlighting the costs of the subsidies and the need to spend both on safety nets for poor segments of society to mitigate the effects of the subsidy removal and on the construction of new refineries and the rehabilitation of existing ones. The SURE brochure summarized the government's case for subsidy removal (Box 5.1), spelled out how much the federal government and states and local governments stood to gain from the subsidy's removal, and announced how the federal government would spend the money saved.

According to the SURE brochure, savings from the removal of the fuel subsidy would be channeled into "a combination of programs to stimulate the economy and alleviate poverty through critical infrastructure and safety net projects." Capital projects would be selected in line with the government's Vision 20:2020 development strategy in the power, roads, transportation, water, and downstream petroleum sectors. The potential impact on the poor of the subsidy's removal would be mitigated "through properly targeted safety net programs." The SURE brochure provided details on the various projects and programs to be undertaken, from the specific road segments to be built to the maternal and child health services to be upgraded.

> ### Box 5.1   *Nigeria: Rationale for Subsidy Removal*
>
> The government summarized its case for subsidy removal in the SURE brochure as follows:
>
> 1. Fixed prices have led to a huge unsustainable subsidy burden.
> 2. Fuel subsidies do not reach intended beneficiaries, and they mostly benefit the rich.
> 3. Subsidy administration has been beset with inefficiencies, leakages, and corruption.
> 4. Subsidy costs have diverted resources away from investment in critical infrastructure.
> 5. Subsidies have discouraged competition and stifled private investment in downstream petroleum.
> 6. Huge price disparity has encouraged smuggling to neighboring countries.

The SURE program envisaged the creation of a specific subsidy savings fund to finance its spending initiatives. The fund itself and the specific spending programs would be overseen by an 18-person board, comprising a chair appointed by the president, only four government representatives, and other members who are respected individuals from a wide cross-section of civil society. The board would seek technical assistance from internationally reputed consulting firms, while an independent body would report to the board directly on implementation.[4]

The government's attempts to win support for its subsidy reform met with strong opposition from powerful sectors of society. In early December 2011, the National Assembly came out against the removal of the gasoline subsidy, claiming that the measure was premature and not supported by firm data underpinning the size and incidence of the subsidies. In response, the Ministry of Finance presented a "Brief on Fuel Subsidies," laying out once again the case for removal, supporting it with data on the explosive growth of the subsidies, and comparing their costs with the government's capital expenditure and borrowing requirements (Okonjo-Iweala, 2011). In addition, several senior officials gave interviews and speeches during the last two weeks of December. However, trade unions were also voicing their strong opposition to the measure, echoing a widely held view that the proceeds from the subsidy removal would most likely go to fund wasteful government spending (including for corrupt politicians) rather than projects to benefit ordinary Nigerians (Okigbo and Enekebe, 2011). State governors who had generally supported the reform earlier on were now silent. Throughout the entire period, the government had deliberately refrained from setting any date for the planned removal of subsidies.

The January 1 announcement came as a surprise and set off widespread protests across the country. On January 9, the two large union federations launched a national strike. Certain parts of the country experienced a near breakdown of

---

[4]President Goodluck officially inaugurated the program on February 13, 2012, and appointed Dr. Christopher Kolade as chair of the SURE Board.

law and order, and there were a number of deaths related to violence and acts of intimidation associated with the strike. On January 15, the president announced that the January 1 price increase would be partly reversed and the new maximum retail price for gasoline would be N97 (US$0.6) per liter, a 40 percent increase over its end-2011 level. However, he emphasized that the government would continue to pursue full deregulation of the downstream gasoline sector. The SURE program would go ahead but would be scaled back in line with the reduced subsidy savings. The president also announced that the legal and regulatory regime for the petroleum industry would be "reviewed to address accountability issues and current lapses." Unions called off their strike that same day.

Mitigating Measures

The SURE program outlined a variety of social safety net programs to mitigate the impact of removing the subsidy on the poor segment of the population. These included

- *Urban mass transit*—Increasing mass transit availability by facilitating the procurement of diesel-run vehicles (subsidized loans, reduced import tariffs, etc.) to established operators. In the first step of this program, the government intended to import 1,600 buses within months.

- *Maternal and child health services*—Expanding the conditional cash transfer program for pregnant women in rural areas and upgrading facilities at clinics.

- *Public works*—Providing temporary employment to youth and women from the poorest populations in environmental projects and maintaining education and health facilities.

- *Vocational training*—Establishing vocational training centers across the country to help tackle the problem of youth unemployment.

Lessons

A well-thought-out public information and consultation campaign is crucial to the success of a reform. Although the government campaigned vigorously for removal of the subsidies, the measure was still highly controversial when it went into effect. The backlash had been predicted. The public communication campaign lasted only six months, and there was no broad popular consultation. The Ministry of Finance produced several short briefs to support its proposal, but these were issued several months into the campaign, and there was no comprehensive report.

The government must establish credibility for its promise that the proceeds from the removal of the subsidy will actually be used for the benefit of the broad population. Notwithstanding the laudable objectives of the SURE program and the plans for oversight by a highly reputable board of directors, the new administration had yet to establish credibility that it would live up to commitments. On the contrary, it suffered from a very negative image of government held by the general public. As such, the subsidy reform was viewed very suspiciously, and the general public simply did not believe that the government would live up to its commitments.

Thorough research on the costs and beneficiaries of subsidies is important to be able to bolster the case for subsidy reform. The absence of good quantitative information on the state of Nigeria's refining industry and of the fuel subsidy mechanism itself allowed spurious arguments, often made by parties with vested interests, that government investment in the state-owned refineries and/or measures to stop abuse by marketers were preferable to removing the subsidies. In addition, the claim that subsidies benefited mostly the poor had been based on anecdotal evidence rather than on research drawing on household survey data.

South Africa

Context

The private sector plays a significant role in South Africa's fuel sector, but prices remain controlled. Six of the seven oil companies[5]—both state-owned and private, including foreign-owned—operate both upstream and downstream in a competitive environment. Some 30 percent of the country's fuel needs are met from synthetic coal-based fuel that is produced domestically, with the remainder derived from imported crude, which is then refined domestically. Although the government has been working toward liberalizing prices, pump prices are currently determined under an automatic pricing mechanism.

Reforms Since the 1950s

The primary reason for introducing the automatic pricing mechanism, which has been in place since the 1950s, was to encourage private sector participation in the energy sector and secure an adequate supply of petroleum products. Concerned about the impact of sanctions on fuel supply during the apartheid era, the government realized that providing prices at least equal to import parity was critical to incentivize international firms to invest and maintain their activities in South

TABLE 5.6

South Africa: Key Macroeconomic Indicators, 1993–2011					
	1993	**1998**	**2003**	**2008**	**2011**
GDP per capita (US$)	3315.6	3100.1	3656.2	5605.8	8078.5
Real GDP growth (percent)	1.2	0.5	2.9	3.6	3.1
Inflation (percent)	9.9	6.9	5.8	11.5	5.0
Public debt (percent GDP)	n.a.	n.a.	36.9	27.4	38.8
Current account balance (percent GDP)	2.1	−1.8	−1.0	−7.2	−3.3
Oil imports (percent GDP)	0.0	0.1	0.1	0.3	0.2
Oil exports (percent GDP)	n.a.	n.a.	0.0	0.0	0.0
Oil consumption per capita (liters)	n.a.	n.a.	441.7	518.2	534.5
Poverty headcount ratio at US$1.25 per day (PPP) (percent of population)	24.3	n.a.	n.a.	n.a.	n.a.

Sources: IEA; IMF, WEO; World Bank, *World Development Indicators*.

---

[5]The six are BP, Caltex, Engen, Sasol, Shell, and Total. PetroSA is the seventh company.

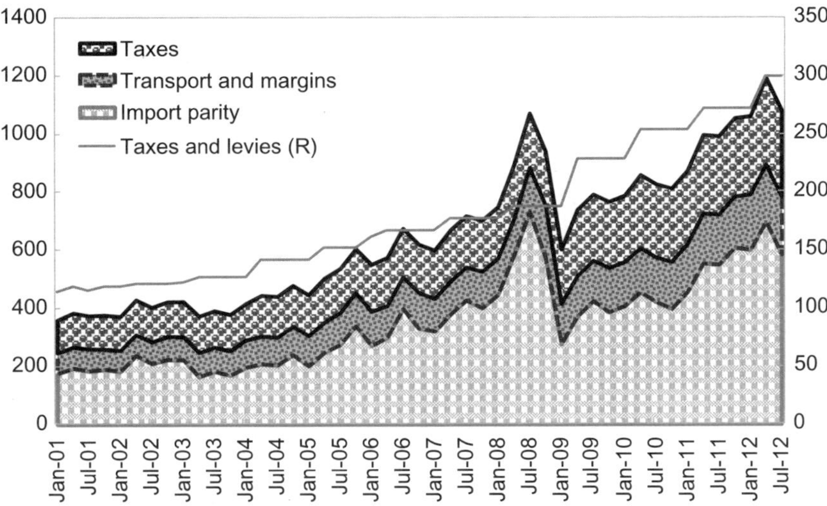

Source: South African Petroleum Industry Association.

Figure 5.7   South Africa: Composition of Gasoline Pump Prices and Taxes, 2001–12 (Cents per liter)

Africa (Competition Tribunal of South Africa, 2006). Most of those international companies remained in South Africa, even during the anti-apartheid embargo.

Attempts to integrate some pump price smoothing through the Equalization Fund over 1977–2004 were not very successful, and they have since been abandoned. The Equalization Fund was established in 1979 and was principally used to smooth out fluctuations in the price of fuel products. It allowed for retail price smoothing by fixing domestic retail fuel prices, with transfers from the fund when international prices were high and transfers to the fund when they were low.[6] When the Equalization Fund went dry, the government needed to finance the deficit. Eventually the government abandoned this policy, and this necessitated substantial increases in prices to bring them to import parity levels. The steep increase in 1993 led to social unrest, which led to the establishment of the Liquid Fuels Industry Task Force to develop a mechanism to address high fuel prices. The current price structure still has the Equalization Fund Levy component, but this has been set at zero since 2002, except when it was used occasionally in early 2003.

The Central Energy Fund, a state-owned entity, was set up in 1977 and given responsibility for determining pump prices on behalf of the Department of Energy.

---

[6]Because domestic prices are adjusted monthly whereas import parity prices change within each month, suppliers can incur deficits or accumulate losses. To address this, the government also introduced "slate" charges into the pricing formula, which could be negative or positive as required. However, these payments have in practice been negligible.

Prices are determined on a monthly basis (the first Wednesday of each month) and include margins, taxes, and levies. The fuel tax—the main tax—is a specific levy announced every February in the budget speech (for implementation in April); it has increased steadily over time, including in periods of rising international prices (Figure 5.7). The decisions of the Central Energy Fund are transparently communicated to the public. There is an online publication[7] of the monthly decisions and price structure, which contributes to a good understanding among the general public of the factors driving pump prices.

Mitigating Measures

No mitigating measures have been introduced in connection with the automatic pricing mechanism. Given the long-standing application of the formula to determine fuel prices, there has been little debate regarding the adverse effects of international fuel price increases.

Lessons

South Africa's success in implementing the automatic price mechanism shows that when the mechanism is well designed, private (including foreign) companies can operate under it without much problem.

The long-standing automatic pricing mechanism has worked well and is likely to remain in place for the foreseeable future. Although South Africa initially implemented the mechanism for strategic reasons under a peculiar political situation, it has been applied consistently over the years. There has been little discussion of an alternative, even when pump prices have had to be increased sharply.

The transparency and credibility of the automatic pricing process has contributed to its durability. South Africa's experience with automatic pricing has been attributed to the credibility that the Central Energy Fund has gained over the years and the transparency with which the mechanism is implemented. The public dissemination of the fund's decisions has contributed to its success.

Stabilization funds can backfire when they are not provided with sufficient resources to absorb volatility in international prices. In South Africa, the Equalization Fund was underfunded, and when resources were exhausted, prices needed to rise sharply—thus defeating the fund's very purpose.

ELECTRICITY SUBSIDIES

Kenya

Context

In line with an expanding economy, Kenya has experienced a substantial increase in energy demand, estimated at 7 percent per year on average over the last six years (Ajodhia, Mulder, and Slot, 2012). Despite improvements in access rates

---

[7] See http://www.energy.gov.za/files/petroleum_frame.html.

TABLE 5.7

Kenya: Key Macroeconomic Indicators, 1995–2009

	1995	2000	2005	2009
Real GDP growth	4.0	2.5	6.1	4.1
CPI Inflation	8.9	8.0	11.1	6.7
Overall balance excluding grants (percent of GDP)	−0.8	−4.1	−4.7	−7.2
Total public debt (percent of GDP)	n.a.	53.1	45.1	44.8
Poverty headcount ratio at US$1.25 per day (PPP) (percent of population)	n.a.	n.a.	43.4	n.a.

Sources: IEA; IMF, WEO; World Bank, *World Development Indicators.*
Note: CPI = Consumer Price Index.

and increases in capacity, electricity generation has not been able to keep up with the increase in demand, and power continues to be a constraint on growth. Kenya depends heavily on hydropower for electricity generation, which accounts for over 56 percent of installed capacity, whereas thermal and geothermal energy sources account for 31 and 13 percent, respectively.

The Kenya Electricity Generating Company (KenGen) is the main player in the wholesale electricity market, accounting for 75 percent of installed capacity as of 2009. It sells power to the retail distributor under several power purchase agreements. In addition, Kenya has five private independent power producers that account for about 25 percent of installed capacity (World Bank, 2010a). The Kenya Power and Lighting Company (KPLC) is responsible for transmission and distribution of electricity. Both KenGen and KPLC operate on a commercial basis and are listed in the Nairobi stock exchange. On the regulatory side, the independent Energy Regulatory Commission (ERC) regulates tariffs, issues licenses, and sets performance targets for KPLC (e.g., revenue collection, average waiting period for new connections, and system losses).

Reforms Since the Mid-1990s

Reform efforts started in the mid-1990s with attempts to rationalize the sector by unbundling electricity generation from transmission and distribution and allowing for private sector participation in the industry. The main objectives of the reform were to improve performance in the power sector, ensure the financial sustainability of the companies operating in the sector, and foster investment. Reform efforts culminated in the 2004 Energy Policy and the 2006 Energy Act. Substantial changes in the tariff structure first occurred in 2005, when revisions were introduced to reflect long-term marginal costs and automatically pass through changes in fuel costs and exchange rate movements. Tariff reform has proved to be durable, but it is important to note that tariff increases occurred concomitantly with improvements in the quality of service. Furthermore, the reform process did not involve any retrenchment of staff in the utilities. The setting up of an Energy Tribunal to arbitrate on disputes between the Energy Regulatory Commission and stakeholders has been instrumental in creating a level playing field in the sector.

Tariffs are based on a formula that, in addition to the basic rate of charge, reflects long-term marginal costs and features a monthly automatic pass-through of generation-related fuel costs and adjustments for exchange rate movements. Furthermore, every six months the formula also takes into account adjustments for domestic inflation. Information on the calculation of tariff adjustments is readily available on the Energy Regulatory Commission's Web site. On the generation side, KenGen has long-term power purchase agreements with KPLC that determine prices and generally reflect underlying costs.

Moreover, residential electricity tariffs in Kenya are based on an increasing block tariff scheme (IBT), such that the unit price per kWh increases according to three defined blocks. The first block ranges from 0 up to 50 kWh per month at a rate of K Sh 2 per kWh. The second block rages from 51 to 1,500 kWh per month at a rate of K Sh 8.10. Finally, the third block applies to households that consume more than 1,500 kWh per month with a rate of 18.57 per kWh. Thus, the tariff rate charged to the highest block is over 828 percent higher than the rate applicable to the lowest one. Residential consumers also pay a fixed charge of K Sh 120. Nonresidential consumers are charged different linear rates, which do not vary according to consumption levels, depending on their category (i.e., commercial, industrial, or government).

Earlier in the reform process, tariff increases faced significant difficulties and required intense negotiations, particularly with large consumers (Bacon, Ley, and Kojima, 2010). Key in securing the cooperation of the private sector was the commitment by the government that the additional cost of energy would help finance the development and expansion of domestic sources of renewable energy that would ultimately reduce the cost of power and strengthen competitiveness. Moreover, there was agreement among stakeholders that ensuring the financial soundness of KenGen and KPLC and setting up a tariff structure reflecting true costs were essential to attract foreign investors into the sector. Subsequently, owing to the negative impact of droughts in 2008 and 2009, a decision was taken to lower the value-added tax (VAT) rate on electricity from 16 to 12 percent.

Power pricing reforms in Kenya allowed tariffs to increase in line with costs from an estimated average of US$0.07 per kWh in 2000 to US$0.15 in 2006 and US$0.19 in 2009 (Table 5.8). The current electricity tariff structure for KPLC tariffs has been in place since July 2008. According to the World Bank (2010a), currently the negotiations for tariff setting and power purchase agreements

TABLE 5.8

Kenya: Key Power Sector Indicators, 1995–2009				
	1995	**2000**	**2005**	**2009**
Access to electricity (percent of population)	11.791	13.102	n.a.	16.10
Electric power consumption (kWh per capita)	130.83	109.72	137.13	147.43
Electric power transmission and distribution losses (percent of output)	17.90	21.16	18.38	15.53
Electricity production (GWh)	3759	4098	5995	6875
Average Tariff (US$/kWh)	n.a.	0.07	0.153	0.19

Sources: Africa Infrastructure Country Diagnostic electricity database; Briceño-Garmendia and Shkaratan (2011a); IMF staff estimates; World Bank (2010a); World Bank, *World Development Indicators*.
Note: GWh = gigawatt-hour.

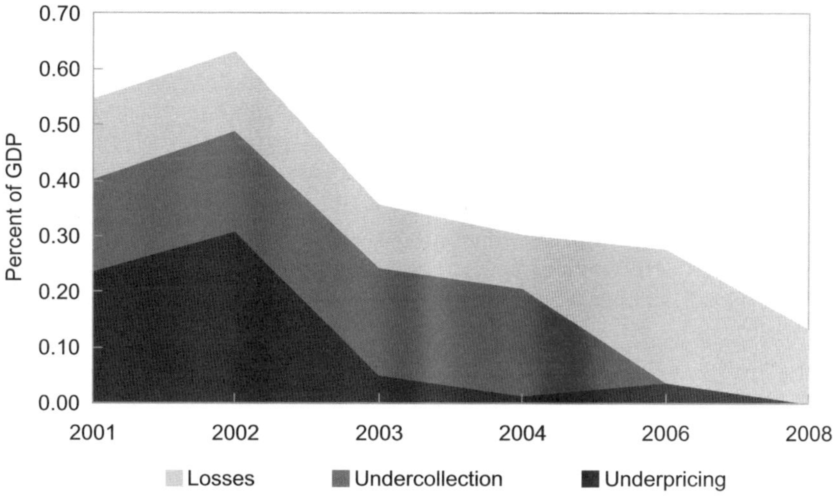

Source: Briceño-Garmendia and Shkaratan (2011a).

Figure 5.8   Kenya: Hidden Costs in the Power Sector, 2001–8
Hidden costs in the power sector have fallen continuously in the last decade.

are transparent; the regulatory framework in the sector is robust and resistant to political interference. However, planned increases in the basic tariff rate in June 2011 did not occur on account of political economy constraints because the authorities believed the prevailing food and energy prices were already excessively high and some delays had been encountered in the implementation of new power generation projects.

As a result of tariff reform measures, the hidden costs of the power sector have decreased significantly over the 2000s, dropping from a level of around 0.6 percent of GDP in 2002 to virtually zero by 2008 (Figure 5.8). In fact, the bulk of the reduction in costs is attributable to large decreases in underpricing, as tariffs were brought to cost-recovery levels, and to reductions in undercollection through improvements in billing. Furthermore, by mid-2008, there were no explicit subsidies or fiscal transfers to power utilities.

The reforms are considered to have been largely successful, with achievements that include rendering both the generation, distribution, and transmission companies financially viable and increasing investment in generation capacity, including some private sector involvement. According to the World Bank (2010a), reforms have resulted in significant operational improvements, including increases in revenue collection. The annual rate of new electricity connections increased from 43,000 in 2003/04 to 200,000 in 2008/09. Distribution losses in the power system also declined gradually from 21 percent in 2000 to 15.5 percent in 2009 (Table 5.8). Revenue collection for KPLC improved from 81 percent in 2004 to 100 percent by 2006 (Foster and Briceño-Garmendia, 2010) before dropping back to about 98 percent, according to the latest information provided by ERC. Labor productivity at KPLC, measured by the ratio of sales per employee or customers per employee, also improved substantially over the period 2004–5 (World Bank, 2010a).

Despite significant progress, there still is a need to expand the power infrastructure to alleviate constraints on growth. The 2007 World Bank Enterprise Survey shows that over 67 percent of firms in Kenya owned a generator and that power outages typically led to losses that amounted to 5 percent of annual sales for the firms surveyed.[8] Briceño-Garmendia and Shkaratan (2011a) present estimates suggesting that unreliable electricity supply reduces Kenya's GDP growth by 1.5 percent per year. Representatives from the Kenya Association of Manufacturers point out that power disruptions continue to affect their operations, despite a provision that prices charged by KPLC to its customers incorporate a requirement that system losses cannot exceed 15 percent.[9]

Mitigating Measures

To address social objectives and affordability concerns, a number of measures have been adopted (World Bank, 2010a; and Briceño-Garmendia and Shkaratan, 2011b). These include

- A rural electrification program that has helped increase the number of connections from 650,000 in 2003 to two million at present;
- A revolving fund, financed by donor funds, for deferred connection fee payments;
- Commercial bank loans for connection fees;
- A lifeline tariff (below costs) for households that consume less than 50 kWh per month, which is cross-subsidized by rates imposed on larger consumers; and
- Cross-subsidies from urban to rural consumers, because tariffs are uniform across these areas.

The 50 kWh per month threshold is commonly used in Africa as a benchmark for the subsistence level of energy consumption. It is estimated to be affordable for 99 percent of Kenyan households (Briceño-Garmendia and Shkaratan, 2011b).

Access continues to be a challenge, particularly in rural areas, where access rates are estimated at 4 percent in 2009 compared with 51 percent for urban locations. Briceño-Garmendia and Shkaratan (2011a) argue that Kenya will need to double its current installed capacity over the next decade and will need to reinforce cross-border transmission links with neighboring countries to increase access to cheaper hydroelectric power and improve overall system security. Despite the fact that there is scope to reduce energy costs through regional interconnections, exchanges across countries in the East Africa power pool are still small.

---

[8] See http://www.enterprisesurveys.org/.

[9] Members of the Kenya Association of Manufacturers account for approximately 60 percent of total industrial energy consumption.

Lessons

Successful electricity reform involves more than tariff changes and it takes time. The reform of the power sector in Kenya started in the mid-1990s and took over 10 years to mature. Apart from a prudent tariff policy, improving the technical and administrative efficiency of state-owned companies was key to eliminating hidden costs. The establishment of a relatively sound regulatory framework, including a regulator that is considered to be largely effective and independent, has also been vital to the durability of the reform process and to encouraging greater private sector participation in generation capacity.

Tariff increases were arguably made more acceptable because they were accompanied by improvements in quality service delivery and access. At the earlier stages of the reform process, authorities actively negotiated changes in tariffs with stakeholders, demonstrating strong political commitment to addressing the challenges of the sector. At the moment, the transparent automatic adjustments to changes in fuel costs (with information regularly published on the Energy Regulatory Commission's Web site), exchange rate movements, and inflation appear to be largely accepted by consumers. Nevertheless, political economy constraints have led to the postponement of a revision in the tariff structure scheduled for mid-2011.

The Kenyan experience also shows that with appropriate instruments, it is possible to reconcile tariff rates at cost-recovery levels with affordability of services for poorer segments of the population. Estimates suggest that the vast majority of Kenyan households are able to afford basic electricity consumption at the effective tariff rate. In addition to the so-called lifeline tariffs (cross-subsidized by large electricity consumers), authorities implemented alternative mechanisms to alleviate the burden of connection fees, such as a revolving fund for deferred payments, financed by donors, as well as commercial bank loans.

Uganda

Context

Despite large potential for hydropower, Uganda has suffered for decades from power shortages. Uganda sustained high economic growth rates during the 1990s and 2000s, which contributed to rapid growth in energy demand (Table 5.9). The public utility, Uganda Electricity Board (UEB), was not able to meet the growing demand partly because of weak financial conditions. Access to electricity was one of the lowest in sub-Saharan Africa, particularly in rural areas. Near exclusive dependence on hydropower prior to 2006 made Uganda vulnerable to weather shocks. As a result of financing constraints, the government was not able to provide adequate support to help UEB meet power demand and tap into the hydropower potential.

In this context, Uganda initiated a comprehensive power sector reform program in 1999. After adopting a power sector restructuring and privatization strategy, a new Electricity Act was passed that aimed at creating an enabling environment for development of the power sector and for private sector participation.

TABLE 5.9

Uganda: Key Macroeconomic and Power Sector Indicators, 2005–10

	2005	2006	2007	2008	2009	2010	2011
Macroeconomic Indicators							
Real GDP growth (in percent)	6.3	10.8	8.4	8.7	7.2	5.2	6.4
Inflation rate (in percent)	10.7	7.2	4.4	12.5	12.3	4.2	15.7
Fiscal balance excl. grants (percent of GDP)	−7.6	−6.1	−6.0	−5.1	−4.8	−7.3	−9.5
Power Sector Indicators							
Input energy (million kWh)	1846	1588	1861	2044	2269	2456	2645
Electricity consumed (million kWh)	1139	1043	1204	1345	1483	1731	1905
Distribution losses (in percent)	38	34	35	34	35	30	28
Collection ratio (percent of all bills)	80	85	93	90	94	96	96
Effective tariff (US cents/kWh)	9	12	18	16	17	16	12
Average cost (US cents/kWh)	13	20	23	26	24	26	26

Sources: IMF, WEO database; Ranganathan and Foster (2012); Uganda, Ministry of Energy and Mineral Development (2012a).

An independent regulatory agency, the Electricity Regulatory Authority, became operational in 2000. In 2001, UEB was unbundled into three separate entities: a generation company (the Uganda Electricity Generation Company Ltd., or UEGCL), a transmission company (the Uganda Electricity Transmission Company, Ltd., or UETCL), and a distribution company (the Uganda Electricity Distribution Company, Ltd., or UEDCL). Given lack of access to electricity in rural areas, a Rural Electrification Agency was established in 2003.

Subsequently, separate private concessions were approved for the generation and distribution companies. In 2003, Eskom Uganda (a subsidiary of Eskom of South Africa) was awarded a 20-year concession for the management of UEGCL's assets. In 2005, UMEME Ltd., the largest electricity distribution company in Uganda, was awarded a 20-year concession for the distribution company UEDCL, the first electricity distribution network concession in sub-Saharan Africa. The state-owned UETCL operates the high-voltage transmission network and serves also as a bulk supplier to the distribution company. As UETCL's bulk supply tariffs have been below cost-recovery levels, the government provided direct and indirect financial supports to UETCL.

The 2005–6 droughts led to an increased dependency on costly thermal power. Prior to the droughts, power generation in Uganda was largely hydro-based. To offset the power shortfall caused by the drought and to meet growing demand, the authorities contracted rental thermal plants, increasing the share of thermal power from about 23 percent in 2006 to about 39 percent in 2011 (Table 5.10). Despite increased thermal power, power cuts were common. According to a 2006 World Bank survey, around 45 percent of firms cited power as a major constraint to doing business (World Bank, 2011a). Despite relying on generators to self-supply for as much as 30 percent of their power needs, these firms lost 10 percent of their sales because of power cuts.

TABLE 5.10

Uganda: Explicit Fiscal Subsidies for the Power Sector and the Cost of Thermal Generation, 2006–11

	2006	2007	2008	2009	2010	2011
Explicit power subsidy						
in US$ million	60.11	51.28	87.56	112.87	151.05	174.80
in percent of GDP	0.6	0.4	0.7	0.8	1.0	1.1
Thermal power (GWh)	370	539	590	896	1022	1029
in percent of total energy	23.3	29.0	28.9	39.5	41.6	38.9
Average oil price per barrel (000 Ush)	131	132	210	132	173	253
percent change (year-on-year)		1	60	−37	32	46
Thermal power costs (in percent of GDP)	0.9	1.1	1.3	1.3	1.5	1.7

Sources: IMF, WEO database; Uganda, Ministry of Energy and Mineral Development (2012b).
Note: Subsidy figures are for fiscal years, which start in July. Data for 2011 are preliminary.

Explicit budgetary support for the power utility has risen steadily since 2005. The explicit subsidy comprised two mechanisms: direct budgetary support to UETCL (bulk supplier) and capacity payments to thermal power units. In FY 2010/11, direct subsidy costs represented 1.1 percent of GDP (Table 5.10). The 2012 tariff increase is expected to eliminate explicit subsidy costs once the Bujagali hydro generation unit becomes fully operational in late 2012. With increased hydro generation capacity, the government will avoid purchase of expensive thermal power, though it will still need to make capacity payments to the independent power producers.

Private concession of the distribution company has produced slow but steady improvements. First, distribution line losses have steadily fallen, from 38 percent in 2005 to 28 percent in 2011 (Table 5.9). Similarly, the collection rate increased from 80 percent of total power bills in 2005 to 96 percent in 2011. To attain these improvements in the distribution system, UMEME invested US$105 million by end-2010—more than envisaged in the contract (Uganda, Ministry of Energy and Mineral Development, 2012a). After little progress in 2005–8, UMEME increased the number of customers by over 30 percent by 2009–10. The increased power supply is expected to further boost the access rate. Notwithstanding this progress, about one-third of the electricity supplied is still not paid for on account of distribution and transmission losses and noncollection of bills.

Once the latter losses are accounted for, we find that the quasi-fiscal deficit (QFD) of the power system has also increased over time.[10] The QFD of the power sector would have amounted to 2.6 percent of Uganda's GDP in 2011—of which about 1.1 percent of GDP were explicit fiscal costs. The QFD continued to grow even after some progress in reducing inefficiencies, largely because of the rising gap between the average effective tariff and the average cost of electricity (Table 5.11).

---

[10]Quasi-fiscal deficit of a power utility is defined as the difference between the actual revenue collected at regulated electricity prices and the revenue required to fully cover the operating costs of production and capital depreciation.

TABLE 5.11

Uganda: Quasi-Fiscal Deficit of the Power Sector, 2005–8 and 2009–11

	2005–8		2009–11	
	In percent of costs*	In percent of GDP	In percent of costs*	In percent of GDP
QFD due to underpricing	32.8	1.0	40.1	1.4
QFD due to distribution losses (up to 10%)	6.7	0.2	6.0	0.2
QFD due to distribution losses (over 10%)	17.0	0.5	12.5	0.4
QFD due to undercollection	4.6	0.1	1.9	0.1
Total quasi-fiscal costs	61.1	1.9	60.5	2.1

Source: Staff calculations based on data from the World Bank; IMF, WEO; and country authorities.
*In percent of total cost of electricity production. QFD = quasi-fiscal deficit.

Growing demand also contributed to the QFD—consumption almost doubled between 2006 and 2011. In any case, QFDs in Uganda have been driven primarily by underpricing: in 2011, underpricing accounted for about 80 percent of the QFD.

Uganda's long-term marginal costs can be substantially lower than the current average costs, but this requires substantial investment. By developing its hydropower potential, the country can reduce costs from US$0.16 to around US$0.12 per kWh (Ranganathan and Foster, 2012). The Bujagali power project was the first step; other major hydro projects are currently being finalized that could double the capacity in a few years.

Reforms Since 2006

Past attempts to bring power tariffs to cost-recovery levels were not enough to catch up with increasing costs. In June and November 2006, power tariffs were increased by about 35 and 41 percent, respectively (World Bank, 2011a). These tariff hikes raised the average effective tariff to US$0.18 per kWh. During 2007–9, no retail tariff adjustments took place, while generation costs kept rising, mainly on account of rising fuel prices, delays in the commissioning of the Bujagali hydropower project, and the depreciating Ugandan shilling (Table 5.11). In January 2010, retail power tariffs were modified to give some relief to household consumers. Given the high cost of thermal power, retail effective tariffs covered only about two-thirds of the costs of power supply in 2010 (World Bank, 2011a).

To offset rising power costs and associated subsidies, the Electricity Regulatory Authority approved a substantial increase in retail tariffs in January 2012. The average effective tariff was increased by about 41 percent (or US$0.05 per kWh). Although at the time of the hike new tariffs were still below the cost-recovery levels, they became broadly in line with the cost recovery when the Bujagali hydropower project became fully operational in October 2012. In addition, the cross-subsidization from households to industrial consumers was also reduced significantly. The new tariff for industrial users, who were previously paying a relatively low price, was set at US$0.13 per kWh—an increase of about 73 percent. The lifeline tariff—for monthly consumption up to 15 kWh—remained unchanged. Following the latest tariff increase, Uganda's power tariffs are in line with those of other members of the East African Community.

Although the recent tariff hike was not without controversies and protests, the government's determination and effective communication have helped to sustain it. The government has run a strong communication campaign to explain the factors that led to the current tariff hike. It was noted that the price of diesel had almost doubled since the last tariff increase in 2006 and that the government was subsidizing consumption as average tariffs remained below unit costs. Although the chair of the Uganda Manufacturers Association pointed out that the new tariff would automatically increase production costs, he also acknowledged that the new tariffs would be bearable if power supply was reliable.

The extent of protests was limited. There were some protests in Kampala and a big political debate in parliament about the tariff hike. The government argued that there were simply no resources to continue subsidizing electricity for the small and relatively rich elite. Low access to power also helped given that the 88 percent of the people without access to electricity were not interested in the protests. Some newspapers highlighted the fact that the subsidy accrues disproportionately to the rich and stressed that the tariff hike would be actually a pro-poor policy decision. What is important is that the lifeline tariff was maintained.

Overall, a variety of factors help create an environment that allowed the authorities to raise power tariffs in early 2012:

- The increasing and unsustainable fiscal costs of thermal power in the context of rising fuel prices: In recent years, the government repeatedly ran arrears in payments for thermal power. In 2011, the explicit fiscal subsidy reached over 1.1 percent of GDP.

- Poorly targeted electricity subsidies: Before the recent tariff hike, large industrial consumers paid less than a quarter of the cost of producing a kilowatt-hour. These consumers accounted for 44 percent of total power consumption in 2010. Thus, almost two-thirds of the power subsidy accrued to a small group of industrial consumers. Among households, only 12 percent of Ugandans have access to the national power grid, whereas the rest rely on unsubsidized kerosene and firewood. The poor generally do not have access to the electricity grid, and the initial power connection costs (about US$80) are too prohibitive.

- Evidence that both industrial and household consumers are willing to pay substantially more than the prevailing tariffs in 2010: A World Bank report noted that average coping costs for intermittent power supply was US$0.3 per kWh (or US$0.4 including fixed costs). For residential customers, the willingness to pay would be US$0.5 per kWh (World Bank, 2011b).

- Investments in hydropower infrastructure leading to a reduction in electricity provision costs over the medium and long term.

- Limited access to power in Uganda: As of 2010, only 12 percent of the population (under 4 percent of the rural population) had access to power, which was less than half of the rate observed on average in other low-income African countries.

Mitigating Measures

The key explicit mitigating measure to power tariff reform is the lifeline tariff for low-income consumers. Uganda has lifeline tariff for poor domestic consumers for power consumption of up to 15 kWh a month. This lifeline tariff has remained unchanged at USh 100 per kWh.

Lessons

The Ugandan case clearly shows that a key impediment to addressing inefficiencies in a power utility is lack of investment. As UMEME made substantial investments, it was able to reduce distribution losses and improve collection while increasing access rates by about 50 percent in the last three years.

Poor financial performance of power utilities is not only caused by the government's desire to maintain low tariffs. Their performance is equally affected by high levels of distribution network losses and undercollection of bills. Therefore, increasing power tariffs alone will not be enough. Power tariffs should be set at economical levels but need to allow for a reasonable level of line losses. In addition, the utility's financial sustainability needs to be pursued through measures to improve efficiency. For this purpose, regulatory policies can help provide utilities with appropriate incentives.

Institutional reform of the power sector takes some time (e.g., 5–10 years). Uganda started its reforms in 1999 and took more than 10 years to make progress in terms of access rates, efficiency measures, and fiscal burden. The reforms led to the establishment of a largely independent regulator with a relatively sound regulatory framework, greater private sector participation in electricity generation and distribution through concessions, and tariff policies that were expected to eliminate hidden costs by the end of 2012.

Tariff increases require a careful strategy for communication and implementation. The Ugandan government communicated well the cost of the power subsidy and its incidence to the public. A large portion of the media considered raising tariffs to be pro-poor measure.

Raising access to power is challenging. Targets for rural electrification had to be revised from 2010 to 2012. It should be noted that the high cost of getting a new power connection is a major impediment to accessing power.

CHAPTER 6

Case Studies from Emerging and Developing Asia

MASAHIRO NOZAKI AND BAOPING SHANG

PETROLEUM PRODUCT SUBSIDIES

Indonesia

Context

Reforming fuel subsidies has been a persistent policy challenge. The size of fuel subsidies in Indonesia has fluctuated considerably over time, reflecting changes in international fuel prices, the exchange rate, and the subsidy regime. The fiscal costs have been generally large, reaching 2.8 percent of GDP in 2008 when international oil prices peaked. In 2011, fuel subsidies were around 2.2 percent of GDP. Indonesia has attempted to tackle subsidy reform a number of times during this period to improve the fiscal position and achieve other policy objectives, such as improving energy efficiency and protecting the environment.

Fuel Pricing Reforms Since 1997

The government cut energy subsidies in the wake of the 1997 Asian financial crisis, and this contributed to political unrest. In the aftermath of that crisis, the government agreed to cut energy subsidies as part of an IMF-supported adjustment program. Instead of the gradual phaseout strategy that was originally envisioned, the government announced increases in the prices of kerosene by 25 percent, of diesel fuel by 60 percent, and of gasoline by 71 percent (Beaton and Lontoh, 2010). The rapid increase triggered protests in the two weeks after the announcement and, along with a complex range of other factors, including dissatisfaction with the government, eventually led to the end of President Suharto's rule.

A number of price increases were implemented between 2000 and 2003 with mixed success and were then rolled back. In 2000, the prices of gasoline, diesel, and kerosene were successfully raised despite violent demonstrations. These prices were again raised in 2001 not only for households but also for industries. An attempt was made in 2003 to automatically link movements in domestic fuel product prices to international prices. This reform, however, was poorly communicated. Many protesters believed that various government decisions at the time had been in favor of powerful interest groups. General dissatisfaction with political corruption and inefficiency also contributed to public opposition. In

TABLE 6.1

Indonesia: Key Macroeconomic Indicators, 2000–2011

	2000	2003	2008	2010	2011
Nominal GDP per capita (US$)	800.0	1091.3	2211.9	2980.8	3508.6
Real GDP growth (percent)	4.2	4.8	6.0	6.2	6.5
Inflation (percent)	3.8	6.8	9.8	5.1	5.4
Overall fiscal balance (percent GDP)	−2.0	−1.4	0.0	−1.2	−1.6
Public debt (percent GDP)	95.1	60.5	33.2	27.4	25.0
Current account balance (percent GDP)	4.8	3.5	0.0	0.8	0.2
Oil imports (percent GDP)	3.5	3.2	4.6	3.4	4.3
Oil exports (percent GDP)	4.8	3.2	3.0	2.2	2.4
Oil consumption per capita (liters)	247.4	254.9	257.8	349.5	n.a.
Poverty headcount ratio at US$1.25 per day (PPP) (percent of population)	n.a.	n.a.	22.6	18.1	n.a.
Fuel subsidies (percent GDP)	n.a.	n.a.	2.8	1.3	2.2

Sources: International Energy Agency (IEA); IMF, *World Economic Outlook* (WEO); World Bank, *World Development Indicators*.
Note: PPP = purchasing power parity.

addition, many of the announced compensation programs did not materialize. As a result, the government rolled back most of the price increases and also severed the link to world prices.

Indonesia became a net oil importer for the first time in 2004 and resumed fuel price increases. Concerned over the increasing fiscal pressure from fuel subsidies, the government undertook two large fuel price increases in 2005. As a result, the price of diesel fuel doubled and that of kerosene nearly tripled. Protests again took place in opposition to the reform, but with less intensity than in 1998 and 2003. The government was led by President Yudhoyono, who was first elected in 2004 and convincingly won reelection in 2009.

Petroleum subsidies have continued, with some reductions. In 2008, with international fuel prices at their peak, petroleum product subsidies reached 2.8 percent of GDP. Fuel prices were raised by 29 percent, on average, and were later reduced as international prices started to fall, though remaining above their pre-increase levels. The government also ceased paying subsidies to larger industrial electricity consumers. The government announced its objective to remove fossil fuel subsidies by 2014. But in September 2010, the House of Representatives agreed to raise budget allocations for subsidized fuel consumption in the revised 2010 budget, which was inconsistent with the government's objective to reduce energy subsidies. Indonesia may have also missed an opportunity to reduce fuel subsides in 2012 as the proposed increases in fuel prices by the government were significantly reduced by the parliament.

The government has begun to encourage liquid petroleum gas (LPG) use over kerosene. Indonesia also initiated a program to phase out the use of kerosene in favor of LPG in 2007. LPG is less subsidized than kerosene and also has lower levels of cost, pollution, and CO_2 emissions. LPG stoves and small LPG cylinders have been distributed free of charge. However, the program was not without problems and may have led to LPG diversion and accidents.

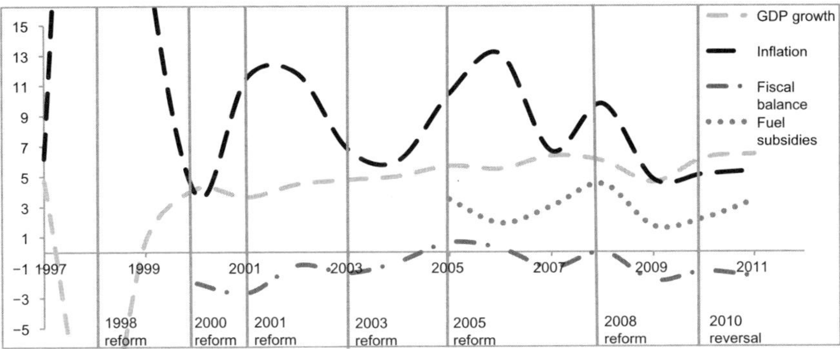

Sources: IMF staff estimates; IMF, WEO database.

Note: 2008 data on fuel subsidy expenditures are based on domestic prices as of mid-2008, instead of end-of-year domestic prices as for other years; in 1998 inflation was 58 percent and real GDP growth was -13 percent.

Figure 6.1 Indonesia: Macroeconomic Developments and Energy Subsidy Reforms, 1997–2011 (Percent of GDP or rate)

Mitigating Measures

Most of the reforms were accompanied by programs to protect the poor. These included the following:

- *Food subsidies, health and education spending, and other social measures.* Subsidies were created for rice; spending was increased on health, education, and social welfare; and support for small business was increased by providing low-interest loans. However, many of the announced compensation programs did not materialize for the reform between 2002 and 2003. In 2008, education support was targeted to the children of the lowest-ranking civil servants, police, and soldiers (Beaton and Lontoh, 2010; Mourougane, 2010).

- *Unconditional cash transfers.* Unconditional cash transfers and other compensating measures were added during the 2005 reform. A number of analyses have credited the reduced intensity of protests in 2005 to the government's decision to compensate poor households for the increase in their living costs through a number of welfare programs. The most high profile program, Bantuan Langsun Tunai, was a series of unconditional monthly cash transfer payments targeted at poor households. The program covered 19.2 million households, or 35 percent of the population, which not only helped the poor but also prevented near-poor households from falling into poverty (Beaton and Lontoh, 2010). Other measures included the health insurance for the poor program, school operational assistance program, and expanded rural infrastructure support project. Awareness of these programs was raised through an extensive public information campaign.

- *Conversion from kerosene to LPG.* An effort to convert households and small businesses from kerosene to LPG has been under way. Kerosene has been widely used in households for cooking and is the most heavily subsidized petroleum product in Indonesia. Reducing the subsidies for kerosene requires an alternative way to provide affordable cooking fuel for households. The same logic also applies to small businesses. In addition to providing a free starter pack, including a stove and a compact cylinder, the government established a communication program to educate the public on the safety of LPG technology. Government statistics indicate that the program has achieved significant savings by increasing the use of LPG and reducing the consumption of kerosene.

Lessons

Targeted cash transfers can reduce opposition to subsidy reform and assist the poor. Indonesia's unconditional cash transfer program was a successful strategy in overcoming social and political opposition to fuel subsidy reforms. Experience with the Bantuan Langsun Tunai program suggests that such programs need good preparation, deployment, and monitoring to effectively assist the poor.

Providing an affordable alternative energy source could also help reduce subsidies and minimize opposition to reforms. Initial data indicate that the kerosene to LPG conversion in Indonesia has been successful. It has achieved the government's goal to reduce fuel subsidies with limited adverse impacts on households and small businesses.

A rapid reduction of subsidies can generate opposition to reform. The sudden, large price increases in 1998 and 2003 were strongly resisted by the public.

Reforms are more likely to be successful with a popular government. The failure of the 1998 reform to some extent reflected public dissatisfaction with the Suharto government. The reforms that followed between 2000 and 2003 were a mix of success and failure, in which the public distrust of the government also played a role. The success of the 2005 and 2008 reforms, in contrast, was helped by the popularity of President Yudhoyono at the time. The erosion of his popularity in recent years, however, may have contributed to the reversal of the reforms.

Reform initiatives are often triggered by adverse economic events, but durable reform requires recognition of the benefits of subsidy removal and long-term commitment to it. The 1998 reform was triggered by the Asian financial crisis. The 2000–2003 reforms were responses to the resulting high fiscal imbalance and government debt. Fiscal pressure and a negative current account balance were the main causes of the 2005 reform as Indonesia became a net oil importer in 2004. And the 2008 reform was the result of historically high oil prices. Without a firm plan for subsidy removal, subsidy reform was stalled in 2010 despite favorable economic conditions.

Ad hoc price adjustments without a clear long-term goal, together with the inability to depoliticize pricing and subsidy policy, led to the reemergence of subsidies and the failure to implement durable reform. Ideally, once the political decision has been made to reduce or remove energy subsides, technical decisions on prices and quantities to subsidize can be delegated to an independent institution that analyzes reform options, disseminates their potential impact, and makes reform recommendations that should be fully implemented. This could improve the transparency of the reform process and reduce the likelihood of setbacks because of election politics. The National Energy Council in Indonesia, however, is not fully independent of the political process. The action by the House of Representatives to increase subsidies in 2010, for example, might have been prevented if there had been a depoliticized decision-making process both for pricing and for the determining quantities to be subsidized.

Communicating the reform objectives and planned mitigating measures to the public can be effective in promoting the acceptance of reforms. As the public becomes better informed about the reasons for, and the objectives of, reforms, it is more likely to understand and accept the measures. Better communicating about the mitigating programs can increase their take-up and thus reduce the negative impact on many households as well as public opposition. The opposition to the 2003 reform in Indonesia was partially motivated by the belief that the reform had been in favor of powerful interest groups.

Philippines

Context

Prior to the deregulation reforms in the late 1990s, the downstream oil sector was heavily regulated, resulting in price subsidies of fuel products when international oil prices rose. The Oil Price Stabilization Fund stabilized domestic prices of fuel products by collecting or paying out the difference between regulated domestic prices and actual import costs. However, increases in domestic prices were politically difficult to implement.[1] As a result, the national government had to replenish the fund by transferring 0.8 percent of GDP in 1990 and 1996.

Reforms Since 1996

The Philippines, a net oil importer, eliminated fuel subsidies through deregulation of the downstream oil industry in the late 1990s. The deregulation largely depoliticized price settings and eliminated fiscal risks by abolishing an oil price stabilization scheme. It was prompted by the government's initiative of economic liberalization; after a "People's Power" revolution in 1986, successive reformist governments abandoned the previous development path of monopoly and protection and liberalized and opened up the economy. At the same time, deregulation was implemented in the context of fiscal consolidation, a key policy objective to restore macroeconomic stability and overcome debt overhang after external debt crises in the early 1980s. These objectives were supported by IMF programs—deregulation of the oil sector was part of the program conditionality of the 1998 Extended Arrangement.

---

[1]For example, an attempt to raise fuel product prices was abandoned because of nationwide protests in 1994. The protest groups consisted of the church, the business sector, labor unions, and other civil society groups (Bernardo and Tang, 2008).

TABLE 6.2

Philippines: Key Macroeconomic Indicators, 2000–2011

	2000	2003	2008	2010	2011
Nominal GDP per capita (US$)	1055.1	1024.8	1918.3	2123.1	2223.4
Real GDP growth (percent)	4.4	5.0	4.2	7.6	3.7
Inflation (percent)	3.8	3.4	8.2	3.8	4.8
Overall fiscal balance (percent GDP)	−3.4	−3.6	0.0	−2.2	−0.8
Public debt (percent GDP)	58.8	68	44.2	42.2	40.5
Current account balance (percent GDP)	−2.7	0.3	2.1	4.5	2.7
Oil imports (percent GDP)	3.9	3.8	12.4	9.6	13.5
Oil exports (percent GDP)	0.4	0.5	1.8	1.4	1.9
Oil consumption per capita (liters)	154.7	150.4	127	140.9	n.a.
Poverty headcount ratio at US$1.25 per day (PPP) (percent of population)	22.5	22	n.a.	n.a.	n.a.

Sources: IEA; IMF, WEO; World Bank, *World Development Indicators.*

The oil deregulation law liberalized the industry and depoliticized the price setting of fuel products. An initial deregulation law was passed in 1996, liberalizing the downstream oil industry and the price setting of fuel products. The Oil Price Stabilization Fund balance improved in 1996 as it received transfers from the national government and fuel prices were raised according to an automatic pricing mechanism after the passage of the deregulation law. The prices were allowed to move freely in February 1997. As a result, the Oil Price Stabilization Fund was abolished, eliminating its cost to the budget (Figure 6.2). When the law abolishing it was ruled unconstitutional by the Supreme Court in November 1997, the government introduced a new law to reinstate deregulation while correcting the constitutional deficiencies of the previous law. The new law was enacted in 1998. The industry remains liberalized today, and movements in international oil prices have been passed through onto domestic prices.

The success of the reform can be attributed to good planning, a well-designed communication strategy, effective consensus building, and strong political will (Bernardo and Tang, 2008). Initially, the political environment was not conducive to such a reform, because President Ramos had won the election by only a small margin and his party was a minority in both chambers of congress. Nevertheless, the reform was planned and communicated soon after the president took office in 1992. A public communication campaign began at an early stage and included a nationwide road show to inform the public of the problems of oil price subsidies. Although the president's party was a minority in congress, he set up a coordination body between the executive and the two chambers of congress and used it to prioritize the oil deregulation bill and forge consensus on it. The commitment of the oil sector reform as part of the conditionality of an IMF program helped to set a time frame for passing legislation. It was opportune that the initial deregulation bill was advanced in 1994–96, a "lull" period with declining inflation, high output growth, and stable exchange rates. Political leadership was exercised when the president pursued the reform despite the Supreme Court ruling that the 1996 deregulation law was unconstitutional—the revised

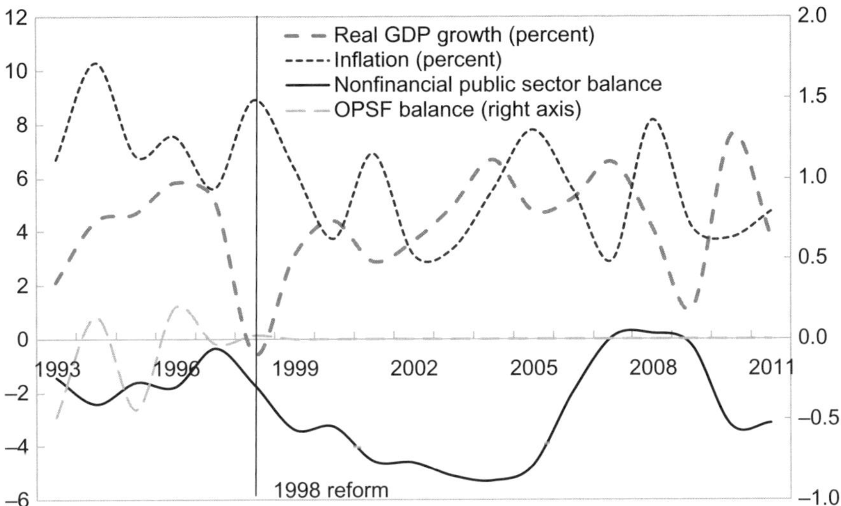

Sources: IMF staff estimates, IMF, WEO.
Note: The Oil Price Stabilization Fund (OPSF) was abolished after the oil sector deregulation in 1998.

Figure 6.2 Philippines: Macroeconomic Developments and Energy Subsidy Reforms, 1993–2011 (Percent of GDP unless otherwise noted)

bill was enacted in 1998 amid a negative growth shock from the Asian crisis, a surge in domestic oil prices resulting from exchange rate depreciation, and renewed political pressure to reregulate the industry.

Mitigating Measures

The authorities introduced appropriate indirect measures to mitigate the effect of the reform (Bernardo and Tang, 2008). For example, the 1996 law included a transition period during which fuel product prices were adjusted monthly using an automatic pricing mechanism. During this period, the government provided transfers to the Oil Price Stabilization Fund to absorb price increases in excess of a threshold. For some years after the deregulation, the government adjusted the duty on oil imports when international oil prices exceeded a threshold, and it used moral suasion to persuade oil companies to adjust prices in small increments.

More recently, the authorities announced several measures to mitigate the impact of the food and fuel crisis in mid-2008. The government launched a package of pro-poor spending programs that are financed by windfall value-added tax (VAT) revenue from high oil prices. The policy package included

- electricity subsidies for indigent families;
- college scholarships for low-income students; and
- subsidized loans to convert engines of public transportation to less costly LPGs (World Bank, 2008).

In addition, the government

- distributed subsidized rice to low-income families; and
- started a pilot project of a conditional cash transfer program in late 2007, which was scaled up in 2008 (Fernandez and Olfindo, 2011).

Lessons

The Philippines' experience with fuel subsidy reform underscores the importance of planning, persistence, and a good communication plan in achieving a successful outcome. The fact that reform efforts started soon after the Ramos administration took office indicates the benefit of advance planning. Commitment to the reform under an IMF-supported program helped set up the time frame for the reform. The reform was supported by a thorough communication strategy, which began at an early stage of the reform. Despite the president's weak political base, a coordination body between the executive and the legislative helped prioritize the reform law. Political leadership was also essential, as evidenced by the government's effort to pass new legislation after the Supreme Court ruling against the first deregulation law.

The survival of the reform to date can be attributed to its comprehensiveness. Rather than ad hoc price adjustments or the simple introduction of an automatic pricing mechanism, the Philippines chose to introduce deeper reforms with liberalization of the downstream oil sector. It succeeded in depoliticizing the price setting of fuel products throughout the product chain, thus raising the bar for a reversal of the reform.

Mitigating measures for the poor during the 2008 fuel price hike helped maintain support for the authorities' approach to fuel pricing. The authorities were able to finance a package of mitigating measures with windfall VAT revenues from higher fuel prices. This was better targeted and a more desirable policy response than a reintroduction of fuel subsidies.

ELECTRICITY SUBSIDIES

Philippines

Context

The Philippine electricity sector became financially unsustainable in the late 1990s. Electricity generation and transmission were monopolized by the state-owned National Power Corporation (NPC) prior to the reform started in 2001. During the 1980s, the NPC's mismanagement led to chronic electricity shortages. To solve this problem, the government opened up the generation sector to independent power producers (IPPs) in the early 1990s to increase supply. Because the NPC was a major purchaser from the IPPs, the IPP initiative left the NPC highly vulnerable to market, exchange rate, and fuel price risks in IPP projects. Eventually, the NPC became financially insolvent in the late 1990s. This was due to the failure to increase tariffs in line with rising costs, as well as the decline in demand (and higher external debt burden) in the aftermath of the 1997 Asian crisis.[2]

Reforms Since 2001

To confront these weaknesses, the government embarked on a major restructuring of the sector in 2001 under the Electric Power Industry Reform Act, which introduced a fundamental reorganization of the electricity sector. It envisaged breaking up the NPC into generation and transmission functions; privatizing generation and transmission assets; unbundling electricity tariffs; establishing the Energy Regulatory Commission, an independent regulatory body that regulates electricity tariffs; creating a wholesale electricity market; and promoting retail competition in the long term. Once it was complete, the reform would eliminate direct fiscal exposure to the electricity sector by depoliticizing tariff setting and limiting government ownership in the electricity sector.

However, passage of the Electric Power Industry Reform Act did not immediately restore the financial sustainability of the electricity sector. Tariff setting remained politicized, despite establishment of the Energy Regulatory Commission in 2001, delaying tariff increases needed to eliminate the operating deficit of the NPC. Limited administrative capacity of the Energy Regulatory Commission was another reason for the delay. In addition, because privatization of generation assets did not pick up until the mid-2000s, the NPC continued to incur losses by purchasing electricity from the IPPs. As a result, the NPC deficit ballooned to 1.5 percent of GDP in 2004 (Figure 6.3).

Substantial tariff increases in 2004–5 were implemented in the context of strong political will to avert a fiscal crisis. The Philippines was on the verge of a fiscal crisis around 2003—the public sector deficit reached 5 percent of GDP on account of weak revenue performance and the large NPC deficit; public debt exceeded 100 percent of GDP and was on a rising path; and deteriorating investor confidence raised external borrowing costs. The Arroyo administration exercised strong political will to avert a crisis and implemented a fiscal consolidation package shortly after the 2004 presidential election. The package included revenue measures, such as increases in VAT and excise tax rates, as well as expenditure restraint. In this context, electricity tariffs were raised by about 30 percent during late 2004 to early 2005, contributing to a reduction in the NPC deficit to 0.2 percent of GDP in 2005 (Figure 6.3). These measures turned out to be successful in averting a fiscal crisis.

Electricity sector reform has continued to advance over the past eight years. In particular, most of the generation assets are now owned by the private sector; wholesale electricity markets are up and running; retail electricity tariffs are unbundled into generation, transmission, and distribution charges; and cross-subsidization among customers has been eliminated except for lifeline tariffs for

---

[2]This study draws on various IMF country reports for the Philippines (IMF, 2005b, 2008; World Bank, 2009).

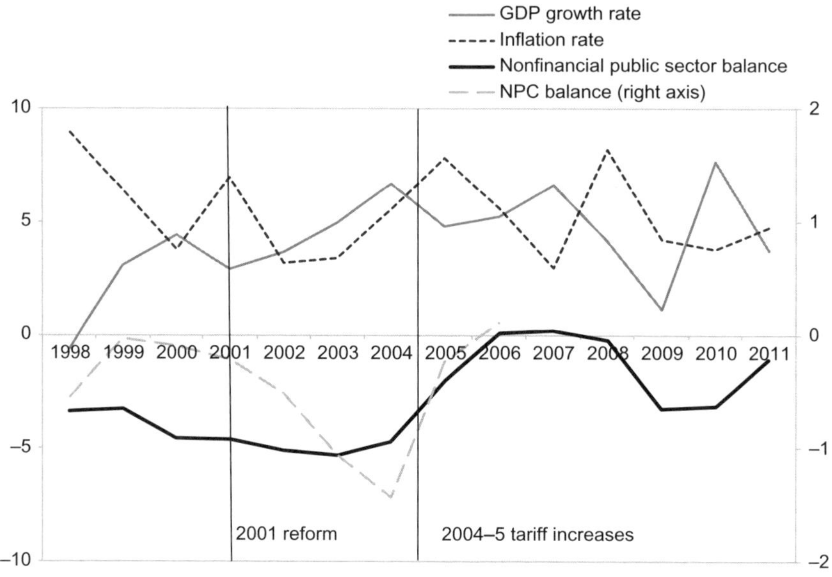

Sources: IMF staff estimates; IMF, WEO.
Note: NPC = National Power Corporation.

Figure 6.3 Philippines: Macroeconomic Developments and Electricity Subsidy Reforms, 1998–2011 (Percent of GDP or rate)

poor families and subsidies provided by the Small Power Utilities Group to users in remote and less developed areas. One remaining issue is settlement of restructuring cost of the NPC—universal electricity charges to cover sunk costs have yet to be introduced.

Mitigating Measures

The adverse effect of the 2004–5 tariff increases on poor households was absorbed mainly by lifeline tariff structures. The Electric Power Industry Reform Act allowed a lifeline tariff schedule as a subsidized rate for poor households. The Energy Regulatory Commission approved lifeline tariff applications for most distributors by 2006. The discount ranged from 5 to 50 percent and benefited three million poor households (Philippines, Department of Energy, 2006).

Lessons

A comprehensive reform that addresses pricing, regulation, and privatization, as well as the mitigation of adverse effects on the poor, can be successful in eliminating electricity subsidies. In the case of the Philippines, establishing an independent electricity regulator helped depoliticize pricing, and privatization reduced the direct fiscal exposure to the electricity sector. A lifeline tariff schedule mitigated the effect of tariff increases on poor households.

Electricity reform can take a while to implement. The Philippines' reform, which started in 2001 and is still ongoing, took a long time to bear fruit. This was because it involves various institutional challenges, such as unbundling the electricity industry, privatizing a large number of generation facilities, and building the capacity of the regulator.

The success of the reform hinged on strong political support throughout the reform process. In the case of the Philippines, the deficit of the state electricity company kept increasing in the early stage of the reform, posing a threat to the country's overall fiscal sustainability. This followed because tariff increases were politicized and delayed even after the establishment of an independent regulator, while privatization of generation assets did not materialize. Tariff increases were eventually approved, because the government remained determined to reduce the deficit of the state electricity company in the context of decisive fiscal consolidation efforts to avert a fiscal crisis.

CHAPTER 7

Case Studies from the Middle East and North Africa Region

Ozgur Demirkol, Luc Moers, and Dragana Ostojic

PETROLEUM PRODUCT SUBSIDIES

Islamic Republic of Iran

Context

Subsidy reform has been on the policy agenda since the late 1980s, with several administrations working on successive reform plans. Setbacks to previous reform efforts led to a surge in energy consumption by the early 2000s, which made Iran one of the countries with the highest energy intensity in the world. As international oil prices approached US$150 per barrel and f.o.b. (free on board) gasoline prices hovered around US$2 per liter, Iran's domestic price of US$0.10 per liter of gasoline was clearly unsustainable. Oil exports were declining, while Iran was importing increasing amounts of gasoline to meet domestic demand, and the relative price differential was fueling smuggling to neighboring countries. The rationing of gasoline, which started in June 2007, reduced demand growth and smuggling to some extent and encouraged development of alternative fuel vehicles, but the price for gasoline purchases in excess of the subsidized quota was set at a still relatively low level of US$0.40 per liter.

Reforms Since 2010

Recognizing the severity of the problems, the authorities launched the first phase of a targeted fuel subsidy reform program in December 2010. The reform made Iran the first major energy-exporting country to drastically cut indirect subsidies and put in place an across-the-board cash transfer program for households. Despite an initial sharp increase in prices, gradual adjustment in prices was a key design feature of the reforms, which planned to increase domestic prices over a five-year period to 90 percent of international prices. In the first phase of the reform, the authorities substantially increased the prices of all major petroleum products and natural gas as well as electricity, water, and bread. In advance of the price adjustments, the authorities also deposited cash transfers in new bank accounts for households, which were to be financed by the revenue from price increases. Part of the revenue from price increases was also allocated to enterprises to help reduce their energy intensity.

TABLE 7.1

Iran: Key Macroeconomic Indicators, 2005–11

	2005	2006	2007	2008	2009	2010	2011
GDP per capita (US$)	2924.6	3428.5	4312.1	4857.1	4926.5	5637.9	6419.6
GDP growth (percent)	4.7	6.2	6.4	0.6	3.9	5.9	2.0
Inflation (percent)	10.4	11.9	18.4	25.4	10.8	12.4	21.5
Overall fiscal balance (percent of GDP)	3.0	2.1	7.4	0.7	1.0	1.6	−0.2
Public debt (percent of GDP)	9.6	8.5	7.8	7.2	8.9	11.3	9.0
Current account balance (percent of GDP)	7.6	8.5	10.6	6.5	2.6	6.0	12.5
Oil imports (precent of GDP)	1.2	2.0	1.9	1.6	1.0	0.4	0.2
Oil exports (percent of GDP)	27.5	26.8	27.5	24.7	19.4	20.7	25.0
Oil consumption per capita (liters)	1,155	1,224	1,217	1,223	1,224	1,108	n.a.
Poverty headcount ratio at US$1.25 per day (PPP) (percent of population)	1.45	n.a.	n.a.	n.a.	n.a.	n.a.	n.a.

Sources: International Energy Agency (IEA); IMF, *World Economic Outlook* (WEO); World Bank, *World Development Indicators*.
Note: Data for 2011 are projected. PPP = purchasing power parity; n.a. = not applicable.

The subsidy reform was also motivated by the authorities' broader structural reform agenda to foster growth and job creation more than to address fiscal concerns. Unlike other countries, Iran's reform was driven by a need to put its valuable hydrocarbon resources to more productive use rather than by a need to reduce the direct burden of subsidies on the fiscal accounts. The Iranian authorities were clear from the outset that the main reform objective was to reduce waste and rationalize consumption. The reform legislation, and the political debate that preceded it, ruled out using the reduction of energy subsidies to improve the country's fiscal balance. The subsidy reform was intended to complement a larger structural reform package that also included financial sector and tax reforms to enhance the competitiveness of the economy.

Despite a good start at the end of 2010, the implementation of the reform program was suspended in late 2012 owing to growing concerns over its financing and the deteriorating macroeconomic situation. In mid-2012 the authorities postponed the implementation of the second phase of the reform because of lack of parliamentary support for the authorities' proposed cash transfer budget and implied price increases under the second phase. Later in November 2012, the parliament formally voted to halt the implementation of the second phase of the subsidy reform, citing rising inflation and unfavorable economic developments in the country. The parliament's vote kept the existing cash transfer program intact but barred further energy price increases under the subsidy reform. The second phase, originally planned for implementation in the second half of 2012, would have involved further increases in energy prices and cash transfers to households. The new round, as originally proposed, was also expected to replace across-the-board cash transfers with more targeted cash transfers for low-income groups.

Mitigating Measures

- *Cash transfers.* About 80 percent of the revenue from price increases was redistributed to households as bimonthly cash transfers. Initially, the authorities leaned toward targeting the transfers to the poorer segments of society. It became clear, however, that it would be administratively difficult to identify and properly screen the recipients given the time line established.

- *Support for enterprise restructuring.* The remaining balance of the revenue from price increases was to be set aside to provide support for enterprise restructuring, with a view to reducing their energy intensity. The authorities conducted a systematic analysis of more than 12,000 enterprises along several criteria to assess the various channels through which the reform could affect them. Out of these enterprises, 7,000 were selected to receive some form of targeted assistance to restructure their operations. This included direct assistance as well as sales of limited quantities of fuels at partially subsidized rates to moderate the impact of the price increase on the input costs of enterprises in the industrial and agricultural sectors.

- *Multitier tariffs differentiated by quantity and region.* Multitier tariffs on electricity, natural gas, and water were used to moderate the impact of the price increases on small users, mostly the poor. Unit tariffs on electricity, natural gas, and water use were set using escalating schedules. Large household consumers were charged prices marginally higher than in international markets. New tariffs also took into account regional disparities in the availability of different heating fuels. Tariff schedules were further differentiated by region, with prices set at lower rates in hot regions with relatively higher air-conditioning demand. Tariff schedules for natural gas and water were similarly differentiated by quantity used and region. In areas where natural gas was not available, heating costs were to remain closely monitored and regulated, and lower-priced kerosene quotas and lifeline electricity rates were provided to ensure affordability of heating.

- *Continuation of gasoline rationing.* The use of the electronic cards system for gasoline rationing and quotas introduced in June 2007 also provided a de facto multitier energy pricing structure for gasoline, making the reform seem gradual. The price of rationed gasoline was increased, but it remained well below the full price at which households could purchase an unlimited amount of fuel. In addition, households were told that they would not lose any of their unused gasoline quotas. Rationing required the implementation of a comprehensive vehicle registration system and personalized distribution and management of the gasoline quotas.

Lessons

Cash transfers to all segments of the population were pivotal in acceptance of the subsidy reform by the population. The authorities initially considered a targeted cash transfer scheme toward the poorer segments of the society but determined that it would be administratively difficult to identify and properly screen the recipients. Also, denying support for the upper-income groups risked triggering public discontent among the biggest energy users. In the end, all citizens were allowed to apply for the compensatory transfers, which were made equal for all applicants. At the same time, with the subsidies being highly regressive, the richest households were encouraged to refrain from applying, with limited success.

Providing all households with equal transfers achieved redistributive effects. For the poor who benefited little from cheap domestic energy prices, the compensation represented a larger share of their income than it did for the middle class; in fact, it was large enough to lift virtually every Iranian out of poverty. In addition, equal transfers helped limit the regressivity of subsidies. This gave the government's economic rationale a powerful public relations stance and built support for the reform.

Maintaining macro-stability has been critical to the success of the reform. Iran suspended the implementation of the second phase of the reform because of concerns over the deteriorating macroeconomic situation. Expansionary monetary and fiscal policies, coupled with the worsening external environment, added to the pressures on the exchange rate, fueled inflation, and put further strain on growth during the implementation of the first phase of the reform. In contrast to the proposed reform, the cash transfer program's budget was reportedly in deficit. Furthermore, high inflation reduced energy prices in real terms and partially offset the impact of energy price increases on consumption, undermining progress under the subsidy reform.

Moving to more energy-efficient production technologies and restructuring enterprises takes more time than initially planned. Although some enterprises were able to continue to expand their production following the subsidy reform, small and medium-sized enterprises were reportedly squeezed by high energy prices and limited government support. There was also reportedly no meaningful progress in adoption of more energy-efficient technologies in enterprises.

Communication is indispensable in creating public ownership of reform. The reform was preceded by an extensive public relations campaign to educate the population on the growing costs of low energy prices and on the benefits expected from the reform. The authorities emphasized that the reform would benefit poor households, which would receive cash benefits, whereas in the past these households had not benefited much from the cheap energy, which was consumed mostly by the richer groups. The Iranian authorities also underlined from the outset that the reforms were not about eliminating subsidies but about switching subsidies from products to households. However, following its implementation, the reform did not seem fully supported by official public information about the de facto implementation and outcome of the reform.

Mauritania

Context

Mauritania's macroeconomic performance since 2000 has been rather volatile (see Table 7.2). GDP growth hovered between −1.2 (2009) and 11.4 percent (2006), while inflation ran between 2.1 (2009) and 12.1 percent (2005). This volatility was partly due to external shocks, partly due to policies. In particular, after the discovery of oil in 2006, the authorities embarked on a fiscal expansion

TABLE 7.2

Mauritania: Key Macroeconomic Indicators, 2000–2010

	2000	2001	2002	2003	2004	2005	2006	2007	2008	2009	2010
GDP per capita (US$)	409.1	412.3	410.5	445.9	504.2	609.5	862.6	878.3	1073.2	897.6	1065.5
GDP growth (percent)	1.9	2.9	1.1	5.6	5.2	5.4	11.4	1.0	3.5	−1.2	5.1
Inflation (percent)	3.3	4.7	3.9	5.2	10.4	12.1	6.2	7.3	7.5	2.1	6.3
Overall fiscal balance (percent of GDP)	0.0	0.0	−2.9	−11.8	−4.8	−7.1	35.8	−1.6	−6.5	−5.1	−1.5
Public debt (percent of GDP)	228.8	223.6	194.5	216.4	209.3	182.1	86.8	96.9	110.5	124.5	86.1
Current account balance (percent of GDP)	−9.0	−11.7	3.0	−13.6	−34.6	−47.2	−1.3	−17.2	−14.8	−10.7	−8.7
Oil imports (percent of GDP)	8.6	7.4	7.4	7.8	9.7	10.6	9.4	15.3	16.5	8.2	9.9
Oil exports (percent of GDP)	0.0	0.0	0.0	0.0	0.0	0.0	0.0	0.0	0.0	0.0	0.0
Fuel consumption per capita (liters)	n.a.	n.a.	n.a.	n.a.	n.a.	359.5	309.3	292.2	294.9	284.2	291.5
Poverty headcount ratio at US$1.25 per day (PPP) (percent of population)	21.2	n.a.	n.a.	n.a.	25.4	n.a.	n.a.	n.a.	23.4	n.a.	n.a.

Sources: IEA; IMF, WEO; World Bank, *World Development Indicators*.

that was reversed only with the start of an IMF-supported program under the Extended Credit Facility (ECF) in March 2010. Mauritania was also hit hard by several droughts and by the 2008–11 spikes in international fuel and food prices.

Reforms Since 2008

Energy subsidy reform in Mauritania was motivated by the above-mentioned fiscal expansion and spikes in international fuel and food prices. The discovery of oil in 2006 prompted large increases in public spending, particularly the wage bill (through adjustment of salaries) and subsidies. When, contrary to expectations, the oil discovery turned out to be very minor, it became clear that the financing of these expenditures was not sustainable, in particular given Mauritania's dependence on volatile mining revenues. The large increases in international fuel and food prices in 2008 and 2011 further increased fiscal pressures. Consequently, subsidy reform, along with wage bill containment, became the cornerstone of the government's fiscal adjustment strategy under the IMF-supported program. The adjustment strategy was designed to free resources while still allowing for much-needed higher social and infrastructure spending.

Better targeting of social protection is an explicit component of the government's fiscal adjustment strategy under the IMF-supported program. The increases in subsidies (diesel, liquid petroleum gas or LPG, electricity) that accompanied the rise in international fuel prices benefited rich households at the expense of the neediest. Almost 80 percent of all energy subsidies were captured by the richest 40 percent of households, thus widening income inequality. Moreover, the emergency relief subsidies on food prices, intended to alleviate the effects of high commodity prices, were not well targeted.

An attempt at energy subsidy reform in 2008 was unsuccessful. A freeze on fuel prices in early 2008 led to huge losses for energy distribution companies (all private). In late June 2008, the government increased the prices of petroleum products by 17.5 percent to 20 percent. No particular public communication strategy was implemented, nor were specific mitigating measures introduced in the context of this fuel subsidy reform episode. Furthermore, subsidy reform–related conditionality was not included in the Poverty Reduction and Growth Facility (PRGF) arrangement covering 2006–9. The one-off price adjustment triggered protests, which contributed to a climate of political instability that culminated in a military coup in August 2008. After the coup, the PRGF was suspended; the price increases were reversed in November 2008.

The timing and magnitude of changes in the prices of petroleum products thus remained discretionary and ad hoc. Prices of petroleum products were controlled by the government and set according to a price structure and formula that in principle was to be adjusted monthly, whenever changes in the international prices or the exchange rate exceeded ±5 percent. In practice, the authorities were reluctant to adjust retail prices. In particular, the government limited price increases when international prices were high (e.g., in 2008), thus causing large losses for distribution companies, and it limited domestic price declines when international prices collapsed, thus allowing petroleum companies to make up for past losses (e.g., in 2009).

The government made progress starting in 2011 under the IMF-supported program agreed after the stabilization of the political situation. The government introduced in May 2012 a new diesel price formula, with the agreement of petroleum distribution companies, following a simplified cost structure. The reform met with relatively limited opposition, despite a price increase of more than 20 percent since January 2011 and the lack of a real public communication strategy. However, unlike in the 2008 episode, the introduction of mitigating measures was an explicit component of the energy subsidy reform strategy. Technical assistance from the IMF fed into the policy dialogue. Despite substantial increases in international fuel prices, the rigorous application of the new simplified automatic fuel price formula on a biweekly basis helped bring domestic fuel prices up to international levels by June 2012 (see Figure 7.1), which was a major achievement.

However, it may still be too early to judge whether gains will prove durable, and much remains to be done. For the rest of 2012, the government was not consistently able to maintain prices at international levels because of the steep

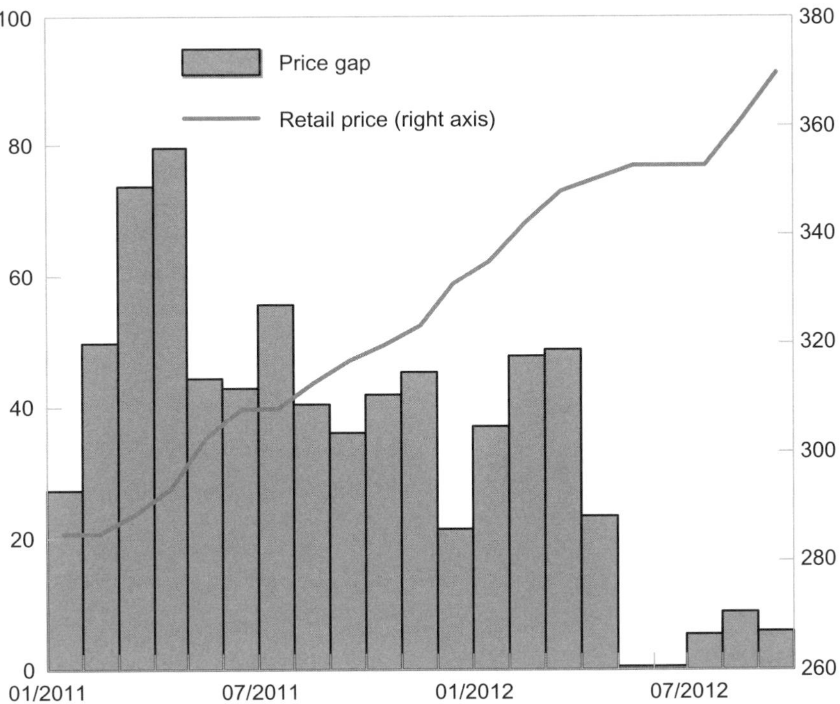

Source: Mauritanian authorities and IMF staff.
Note: The full pass-through price is calculated as the sum of import cost, margins, and taxes. Price gap is calculated as the difference between full pass-through price and domestic retail price.

Figure 7.1 Mauritania: Diesel Retail Price and Price Gap, 2011–12 (Ouguiyas per liter)

increase in world prices. To ensure that the pricing formula can continue to be applied automatically even in the face of sharp fluctuations in international prices, the government intends to introduce a cap of 3 percent on any one adjustment in cases when the formula would dictate a bigger change. This smoothing approach should avoid excessive domestic retail price volatility, which could undermine the political support for the formula. Additional reduction in subsidies will follow planned increases in electricity tariffs (for large consumers) and in gas prices.

High international prices also aggravated the cost of subsidies to the electricity sector. SOMELEC, the public electricity company that produces almost all the electricity in Mauritania, incurred significant losses from the increase in international fuel prices. Two-thirds of the electricity consumed in the country is generated using thermal plants, evenly split between diesel and fuel oil. Despite higher international prices, electricity tariffs have not been revised upward in recent years. Residential and commercial tariffs are among the lowest in the region and are estimated at more than 30 percent below cost-recovery prices.

Supported by the IMF-supported program, the government also moved to address the electricity subsidies. A restructuring plan was laid out with the help of the World Bank and the French Development Agency. The government recapitalized SOMELEC and clarified its financial relationship with it by (1) paying its electricity bills on time; (2) providing SOMELEC with the required subsidy for its operations at regular intervals throughout the year; and (3) drawing out a plan for the settlement of arrears accumulated through the end of 2010. Furthermore, electricity rates for the services sector were aligned with the rates for medium-voltage electricity starting at the beginning of 2012. These measures, together with a new credit line from the Islamic Development Bank, enabled the company to significantly limit its recourse to bank borrowing at high interest rates, which had been a drain on its finances in the past. A tariff study, conducted by an international firm, will be completed in 2013 and will result in electricity rates being increased, particularly those paid by large consumers. In addition, the authorities have called on a consulting firm to establish a performance contract between SOMELEC and the government.

Mitigating Measures

- *Emergency measures.* In 2011, the Mauritanian authorities introduced emergency relief measures to mitigate the impact on the poor of higher international fuel prices and a drought, which led to a food emergency. Unlike the 2008 emergency plan, the new package, which was worth about UM 40 billion (3.4 percent of GDP) and was the largest in terms of GDP among the region's oil importers, comprised mostly reversible measures (e.g., it did not include a raise in the wages of civil servants). It was thus an improvement over earlier measures, and some social response by the government was certainly needed.

- *Plans for the social safety net.* However, the IMF-supported government program envisages substituting this temporary program with permanent well-targeted social safety nets. The government plans to conduct a full assessment of the existing drought-emergency program, particularly the functioning of the "subsidized-food shops" programs, which were extended through the end of 2012. This food subsidy program has not always been effective in reaching the poorest households in rural areas. Moreover, with the worst of the drought's impact now over, there is an opportunity to gradually remove most components of this emergency program, reorienting the savings toward scaling up well-targeted cash transfer schemes.

- *Cash transfers.* With the assistance of the World Food Program, a start has been made with such a cash transfer program. This program, which was rapidly put in place, targets 10,000 vulnerable households in Nouakchott identified through a recent poverty survey. Each household receives UM 15,000 monthly (equivalent to half of the legal minimum wage) via a bank transfer. A positive side effect is that beneficiaries thus also gain access to financial services. The program was extended in June 2012 to 15,000 households in four rural areas deemed to have high food insecurity. The agenda of scaling up such well-targeted cash transfer schemes should benefit from the expansion of the vulnerability and poverty survey to provide nationwide coverage, because most of the poor are in rural areas.

- *Broader social protection.* A broader social protection strategy developed with UNICEF will also further strengthen the coverage of the social protection system and better protect the poor and vulnerable. Accordingly, with the assistance of technical and financial partners, the authorities plan to strengthen programs such as free school cafeterias, food-for-work, and support for pregnant women. Moreover, recognizing the adverse effects of drought on food security, they are developing a national food security strategy for the period 2015–30 and an associated national investment program.

Lessons

Depoliticizing fuel price adjustments as much as possible can help lock in initial price gains. The automatic implementation of Mauritania's new diesel-price formula has been very effective in keeping a lid on subsidies. Putting a cap on any one price change would ensure that large international price fluctuations do not lead to excessive retail price volatility, which could undermine political support for the automatic fuel price formula. At the same time, such price smoothing would still allow domestic prices to follow the trend in international prices.

Too rapid a reduction of subsidies can generate opposition to reforms. The sudden, large price increases in 2008 met with strong opposition, stimulated political instability, and ultimately had to be rolled back. The absence of any mitigating social measures at the time exacerbated the situation.

Mitigating social measures can help address opposition to energy price increases and their impact on the poor, but they should be well targeted. Mauritania's recent cash-transfer schemes, developed with assistance from the World Food Program, appear promising in this respect. In contrast, the earlier emergency-relief programs were less well targeted and not as effective. Furthermore, care should be taken that temporary emergency programs do not become permanent entitlements, draining fiscal resources. The absence of a fully fledged communication campaign has not been an obstacle to reforms in Mauritania so far. However, the authorities are well advised to accompany energy subsidy reform with an explicit communication campaign that explains its benefits to the population. Transparent reporting on the use of freed-up budget resources should also increase public confidence in the outcome of the reform.

The linkages between fuel and electricity subsidy reform need to be explicitly recognized and addressed. If a highly subsidized electricity sector uses large amounts of fuel, as in Mauritania, fuel price increases can add to problems in the electricity sector. In the case of public sector electricity utilities, reform should also be accompanied by the clarification of their financial relationship with the government.

Involving donor partners that have specialized in other areas can increase the reforms' chance of success. In the Mauritanian case, the role of the World Food Program and UNICEF in the development of social mitigation strategies was clearly helpful. The study on the restructuring of the electricity sector and SOMELEC assisted by the World Bank and the French Development Agency was key in addressing electricity subsidies.

Yemen

Context

Yemen has undertaken several reforms to reduce fuel subsidies since the 1990s. The size of these subsidies has fluctuated over time, reflecting changes in international fuel prices, consumption volumes, the exchange rate, and domestic prices. Yemen's main goal in subsidy reform has been to improve its fiscal position, while paying due attention to social considerations. Despite these reforms, the subsidy bill in the budget remains large, around 10 percent of GDP in 2012 (having peaked at 14 percent of GDP in 2008). This amount exceeds the total of infrastructure and social expenditures.

Reforms Since 1994

After 1994, the government increased the price of gasoline, but currency depreciation wiped out all the gains from domestic price increases. In the aftermath of the 1990–94 civil war, the government increased the price of fuel products, which is consumed more by better-off households, by 75 percent. However, the depreciation of domestic currency of almost 240 percent that took place in 1995 wiped out the gains from this increase. In 1995–96 the government implemented more price increases, which affected four products. Gasoline increased by 80 percent, diesel by 100 percent, and kerosene by 189 percent; LPG increased in two steps (first by 123 percent and then by 85 percent). However, prices in dollar terms remained well below their 1994 levels. During 2000–2004, the government increased the price of diesel again by 30 percent in two consecutive years.

TABLE 7.3

Yemen: Key Macroeconomic Indicators, 2000–2011

	2000	2001	2002	2003	2004	2005	2006	2007	2008	2009	2010	2011
GDP per capita (US$)	539.6	532.4	560.0	597.8	682.1	797.7	881.6	971.3	1171.1	1061.0	1272.5	1343.3
GDP growth (percent)	6.2	3.8	3.9	3.7	4.0	5.6	3.2	3.3	3.6	3.9	7.7	−10.5
Inflation (percent)	12.2	11.9	12.2	10.8	12.5	9.9	10.8	7.9	19.0	3.7	11.2	19.5
Overall fiscal balance (percent of GDP)	6.1	2.8	−0.6	−4.2	−2.2	−1.8	1.2	−7.2	−4.5	−10.2	−4.0	−4.3
Public debt (percent of GDP)	61.2	60.7	57.8	56.8	52.1	43.8	40.8	40.4	36.4	49.8	40.9	42.4
Current account balance (percent of GDP)	13.8	6.8	4.1	1.5	1.6	3.8	1.1	−7.0	−4.6	−10.2	−4.4	−3.0
Oil imports (precent of GDP)	2.2	5.1	6.2	6.8	7.5	10.5	17.7	18.1	13.3	7.8	6.7	8.7
Oil exports (percent of GDP)	35.1	29.5	29.4	29.3	31.0	35.6	35.3	28.3	28.7	17.6	20.2	23.3
Oil consumption per capita (liters)	n.a.	252.3	260.8	265.4	274.4	280.5	279.1	293.8	305.4	316.4	322.6	n.a.

Source: IEA; IMF, WEO; World Bank, *World Development Indicators*.

Nonetheless, in dollar terms, the price of diesel remained below its level of a decade earlier. Throughout the 1994–2004 period, the depreciation of the currency wiped out all the gains from domestic price increases. During this time, the government also discouraged the use of kerosene for cooking by making it more expensive compared with LPG.

The most important subsidy reform, launched in 2005, aimed at gradually adjusting domestic prices over the medium term. This reform was based on a World Bank study and IMF policy advice, which underlined the need to preserve fiscal sustainability in the face of declining oil reserves. So, in July 2005, the government increased domestic prices by 130 percent on average. This price increase led to violent protests, and the government reacted by partially reversing it. Nonetheless, the net price adjustment remained substantial, at 71 percent for gasoline, 106 percent for diesel, 119 percent for kerosene, and 7 percent for LPG. There was no increase in the price of mazot, which is used mainly for electricity generation.[1] It is important to note that social tensions during this episode were related not only to the subsidy reform but also to reforms in the taxation system. The initial relative success of the fuel price adjustments was canceled by the spike in commodity prices in later years. As a result, the subsidy bill remained high, at almost 9 percent of GDP in 2005.

In 2010, as a part of the reforms supported by an IMF Extended Credit Facility arrangement, the prices of gasoline, diesel, and kerosene were gradually increased by about 30 percent on average, and the price of LPG was doubled over a period of nine months. The reform strategy was based on technical assistance from the World Bank, which drew lessons from the experience of the previous reforms. However, the public information campaign component of the strategy was not adopted. Instead, the government implemented small surprise increases. In addition to fuel price increases, the government also introduced some efficiency-promoting measures, such as replacing diesel-fueled power generators with gas-fueled ones. In late 2010, Yemen started to differentiate diesel prices by charging higher prices to commercial users. The main objective of this reform episode was to reduce fiscal pressures, following the record-high fiscal deficit of 10 percent of GDP in 2009.

In 2011–12, as a consequence of the political crisis and tight fiscal space, the government increased the price of gasoline by 66 percent and doubled the prices of diesel and kerosene.[2] Overall, this reform episode was accepted by the population despite the political tension between the ruling party and the opposition. The major pipeline that provides oil to domestic refineries had been sabotaged. At the same time, the government was able to import only limited quantities of refined fuel products. The ensuing fuel scarcity resulted in the emergence of a black market, with prices that were a multiple of the official sale prices, and long

---

[1] In addition, the electricity company receives diesel at a lower price than other users.

[2] The price of gasoline was initially increased by 133 percent for 90 percent of consumers, and for 10 percent of consumers (poor households who use gasoline) it was left unchanged. In 2011 the increase was partially reversed, but prices were unified.

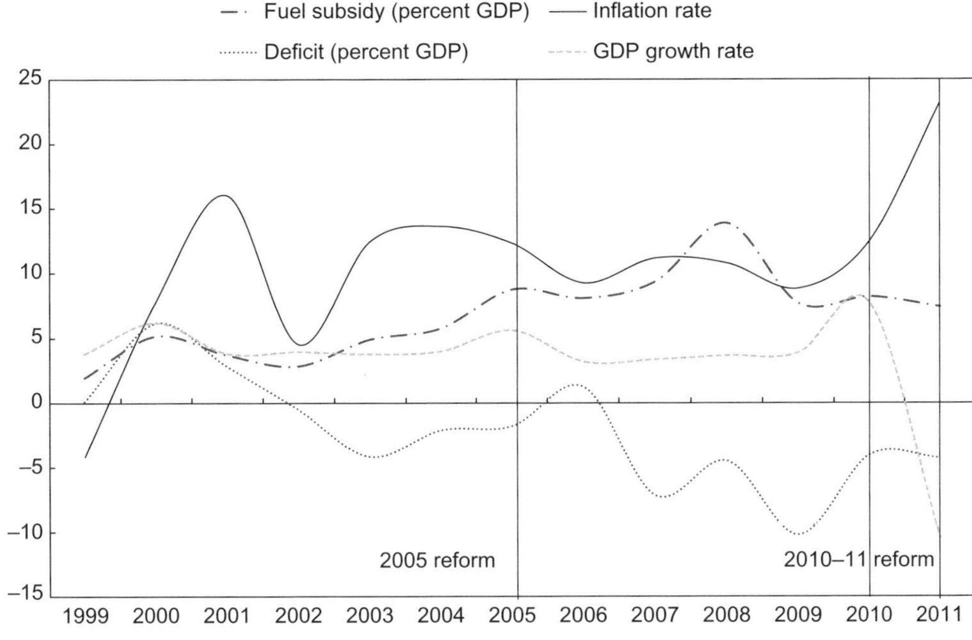

Sources: IMF staff and authorities.

Figure 7.2 Yemen: Macroeconomic Developments and Energy Subsidy Reforms, 1999–2011 (Percent of GDP or rate)

lines at the gas stations. This situation may have played a role in the population's acceptance of the official price increases in exchange for the benefit of uninterrupted supply.

Mitigating Measures

Well-off Yemeni households benefit disproportionately from fuel price subsidies, both directly (because they consume more energy than do poorer households) and indirectly (because they consume more energy-intensive products and services). Overall, about 40 percent of fuel subsidies go to the richest 20 percent of households, while only 25 percent goes to households in the bottom 40 percent (based on updated 2005 Household Budget Survey data). The unequal distribution of benefits varies widely by fuel product. In the case of gasoline, for example, households in the bottom 40 percent receive only 10 percent of the direct value of the subsidy.

To mitigate the impact of the past subsidy reforms on the poor, the authorities introduced or strengthened the following components of the social safety net:

- *Conditional cash transfers.* The Social Welfare Fund was established in 1996 as a poverty alleviation program to provide conditional cash transfers to households. The coverage of the fund was expanded gradually, and transfers were increased in steps. The transfers were meant to partly mitigate the impact of fuel subsidy reforms. The timeliness in the implementation of measures addressing social support varied. For example, in the 2005 subsidy reform episode, it took three years to approve a social protection law to allow for more streamlined application for benefits and increase monthly transfers. Conversely, the 2010 reform was almost simultaneously mitigated by a 50 percent expansion of the coverage of the cash transfer scheme. Thus far, there have been no mitigating measures in the 2011–12 reform episode, but the government is considering a further increase in the Social Welfare Fund coverage or the size of existing transfers.

- *Public works.* A program focusing primarily on poverty prevention, the Public Works Project provides short-term employment and support for small-scale contractors through a labor-intensive public works program.

- *Community and enterprise development.* In addition, the Social Fund for Development promotes community and small- and micro-enterprise development and provides short-term employment for both the transitory and chronically poor.

- *Fuel conversion.* Other mitigating measures include conversion from more to less expensive fuels. For example, the government promoted the conversion from kerosene to LPG for residential use starting in the early 2000s. Also, in 2010 the diesel-fueled electricity plants were converted to natural gas.

Lessons

The experience with the pace and size of price increases varied. When the public was made aware of the need for and the benefits of reforms (e.g., to ensure adequate supply), it accepted large adjustment in prices. Conversely, when reforms were not accompanied by implementation of an effective public information strategy, especially at times of heightened political tension, there were popular protests that forced at least a partial reversal of the adjustments. Adequate planning to strengthen the safety net and communicate assurances regarding the mitigating effort was also essential for gaining public support.

It is important to avoid multiple prices for a single fuel product. The Yemeni government introduced multiple pricing for gasoline and diesel with a view to protecting vulnerable households. However, the implementation of that strategy was difficult and gave rise to arbitrage and distortions. It also raised a governance challenge because it created an incentive for commercial users to try to obtain the product at the cheaper price intended for residential users. This is not the case for metered products, such as electricity, for which differential pricing is easier to implement.

If well designed and implemented in a timely fashion, cash transfers and other social protection programs can be effective in protecting the poor and reducing opposition to reforms. The Social Welfare Fund cash transfers, as well as support from the Public Works Project and the Social Fund for Development, helped to reduce opposition to reforms. Based on this experience, it is possible to argue that

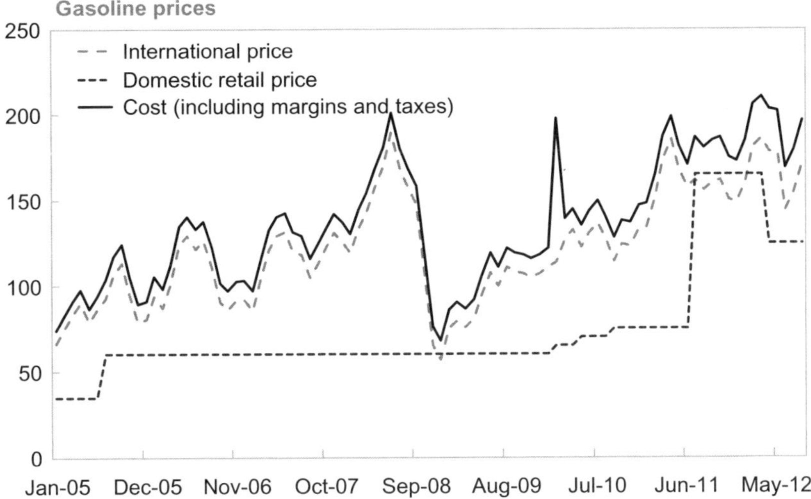

Sources: IMF staff and authorities.

Figure 7.3   Yemen: Fuel Prices and International Full Pass-through Prices, 2005–12 (Yemini rials per liter)

opposition to the 2005 reforms could have been reduced if the social protection programs had been simultaneously implemented.

Although adverse economic conditions increase the need for reforms, they can make price adjustments more difficult, especially if combined with political tensions. It is therefore important whenever possible to implement reforms in a timely fashion before economic and social conditions deteriorate further. The large resources used in generalized subsidies can then be used more effectively to target the poor and to spur growth and employment creation.

Efficiency and governance improvements can also help reduce subsidy costs. Adequate relative pricing (e.g., natural gas vs. diesel and LPG vs. kerosene) provides an incentive for efficient switching of consumption. Strengthening governance would also help to enhance targeting and reduce abuse and smuggling.

CHAPTER 8

Case Studies from the Latin America and Caribbean Region

ALLAN DIZIOLI, JAVIER KAPSOLI, MASAHIRO NOZAKI, AND MAURICIO SOTO

PETROLEUM PRODUCT SUBSIDIES

Brazil

Context

Brazil's economic performance in the prereform era of the 1980s was characterized by low growth, high inflation, and substantial fiscal imbalances. Economic growth averaged about 3 percent and inflation averaged 272 percent. Fiscal policy was expansionary, with the overall budget deficit averaging 5 percent of GDP during the period and reaching 7 percent of GDP in 1989. Weak fiscal performance led to an increase in net public debt from 24 percent of GDP in 1981 to almost 40 percent of GDP in 1989. These deteriorating conditions put pressure on the authorities to alter Brazil's import-substitution policies and liberalize the economy (Giambiagi and Moreira, 1999), including in the energy sector.

The state-owned oil company, Petrobras, dominated the oil market in the 1980s. It held a monopoly on the upstream market and on the refining of liquid fuels in Brazil. In addition, Petrobras had a monopoly on crude oil and petroleum product imports. Even though the distribution of fuel products was open to private sector companies, including multinationals, the final consumer price was determined by the government. An oil stabilization fund was established in 1980 to smooth crude oil price volatility. The price of oil sold to the refineries was adjusted to keep the oil costs for Petrobras refineries at a set price determined by the government; the fund accumulated contingent liabilities to Petrobras when international crude prices were high, and these were offset when crude prices were low. The prices established for diesel and liquefied petroleum gas (LPG) were also consistently set below import-parity costs. As a result of increasing oil import costs, the crude oil stabilization fund and Petrobras ran up enormous deficits. To pay for these accumulated losses, the government transferred R$5.8 billion (0.8 percent of 1995 GDP) to Petrobras in the mid-1990s, and Petrobras had to absorb other losses that were never transparently recorded on the budget.

TABLE 8.1

Brazil: Key Macroeconomic Indicators, 2000–2011

	2000	2003	2008	2010	2011
GDP per capita (US$)	3751	3104	8729	10816	12917
GDP growth (percent)	4.31	1.15	5.16	7.49	3.77
Inflation (percent)	6.18	13.72	8.33	8.23	6.97
Overall fiscal balance (percent of GDP)	−3.37	−5.31	−2.34	−5.93	−3.57
Gross public debt (percent of GDP)	51.1	59.6	58.5	63.7	62.2
Net public debt (percent of GDP)	47.7	54.9	38.1	40.2	38.6
Current account balance (percent of GDP)	−3.76	0.76	−1.71	−2.21	−2.12
Oil imports (percent of GDP)	1.19	1.16	1.84	1.21	1.25
Oil exports (percent of GDP)	0.16	0.33	0.33	0.19	0.23
Oil consumption per capita (liters)	412	394	482	624	n.a.
Poverty headcount ratio at US$1.25 per day (PPP) (percent of population)	11.82	11.21	6.01	6.14	n.a.

Sources: International Energy Agency; IMF, *World Economic Outlook* (WEO); World Bank, *World Development Indicators*.
Note: PPP = purchasing power parity; n.a. = not applicable.

Fuel Pricing Reforms—Early 1990s to 2001

A gradual approach to the removal of subsidies was chosen by the government to deal with opposition from interest groups. To build public support for the reforms, the government promised consumers that privatization and liberalization would lower energy prices and improve services. Even though low prices to consumers had led to the subsidies, the authorities hoped that improvements in efficiency of the refinery would be sufficient to reduce these outlays without increases in consumer prices.

There were several steps involved in liberalizing fuel prices. The process of liberalizing the market began in the early 1990s with the liberalization of prices for petroleum products used primarily by firms, such as asphalt and lubricants (see Figure 8.1). This was followed by a more extensive liberalization that included gasoline prices for final consumers in 1996, LPG for final consumers in 1998, and diesel in 2001. The first products to lose subsidies were generally those consumed by politically weak stakeholders, whereas the politically more difficult subsidies (for liquid fuels used for transport and industry) were removed later. The removal of subsidies for ethanol producers and the suppliers of equipment and services to Petrobras was left to the end of the liberalization program.

Price liberalizations were associated with short-term increases in inflation. The dynamic effects of the liberalization reforms can be seen in Figure 8.1. After each reform, there was a spike in inflation in the short term that eventually died out over the longer term as prices were allowed to fluctuate with developments in international markets.

Petrobras maintained a dominant role in the market despite liberalization. In 1995, the formal monopoly of Petrobras on the upstream market, on refining liquid fuels, and on the imports of crude oil was revoked. In 1997, the Agência Nacional do Petróleo was created to oversee the deregulation and restructuring of the sector and to manage the auctioning of oil fields for exploration. Despite

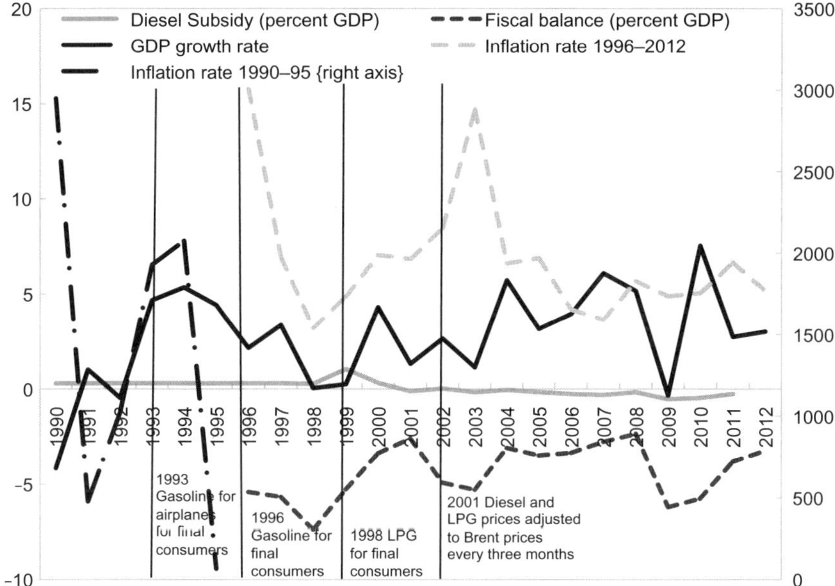

Sources: IMF staff and authorities.
Note: LPG = liquefied petroleum gas.

Figure 8.1   Brazil: Macroeconomic Developments and Energy Subsidy Reforms, 1990–2012 (Percent of GDP or rate)

the wide-ranging scope of the authorities' privatization efforts, Petrobras has still managed to preserve a de facto monopoly in refining and distribution.

High rates of inflation and currency depreciation posed significant challenges for containing the fiscal costs of subsidies. To avoid the emergence of subsidies, frequent price increases were necessary in an environment of high inflation. However, diesel price increases did not keep pace with exchange rate depreciation in the late 1990s, leading to an upward spike in diesel subsidies to about 1 percent of GDP in 1999 (Figure 8.1).

Fuel Price Setting Since 2002

Official price liberalization for all fuel products has been in effect since 2002, and this has helped prevent the recurrence of subsidies. Prices were increased and remained above international levels, despite significant pressure on the currency between 2001 and 2003. Fuel prices continued to rise steadily until 2005 and after that remained mostly flat despite fluctuations in international prices (Figure 8.2). There is no official government price setting in the chain of fuel production and marketing. Under the new regulatory scheme, the Agência Nacional do Petróleo monitors fuel prices through its "survey of fuel prices and margins," which includes gasoline, fuel ethanol, diesel, natural gas for vehicles, and liquefied natural gas.

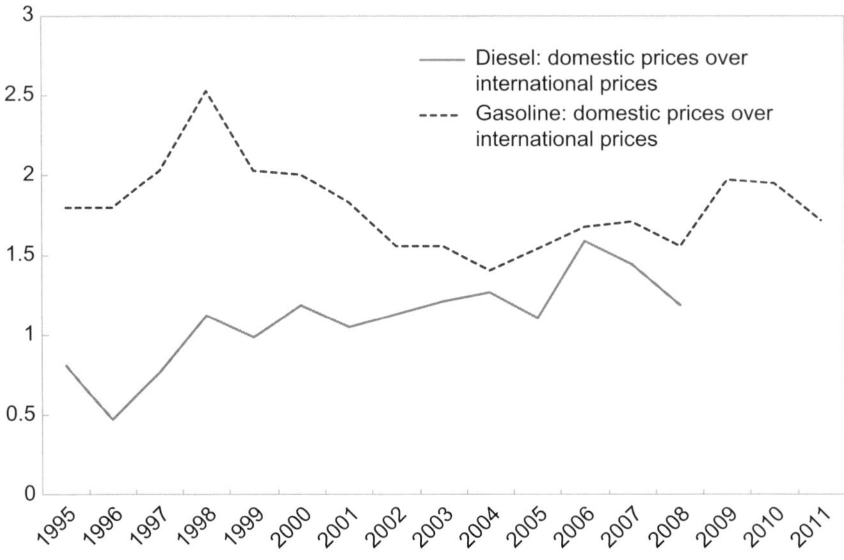

Sources: Country authorities and IMF staff estimates.

Figure 8.2 Brazil: Fuel Price Developments, 1995–2011

Durability of the Reforms

Although officially oil prices are determined by Petrobras, in practice the government has used prices as a tool to control inflation. For instance, the government reduced taxes on gasoline and diesel in 2004 and removed the taxes on LPG and fuel oil so as to keep petroleum prices constant for final consumers. As a result of the lower levy rate and narrower coverage, the aggregate total amount of petroleum taxes has not increased in spite of growing consumption. In most years, the price received by Petrobras, net of fuel taxes, has been insufficient to cover its costs on the sale of these products. The operational losses on this downstream business have been offset by profits on Petrobras's upstream operations.

More recently, the incomplete pass-through of changes in import prices has also had consequences for Petrobras's profits. With the increase in international fuel prices and the sharp exchange rate depreciation between 2010 and 2012, fuel import prices rose substantially and were not fully passed on to consumers. The ratio of domestic prices to international prices for gasoline fell from 1.95 to 1.47 between 2010 and 2012, and the ratio for diesel fell from 1.44 to 1.02. Despite the fuel tax cuts, the net prices received by Petrobras were not enough to compensate for rising import prices. In 2012, Petrobras had profits of US$10 billion, the lowest level since 2004 and a reduction of 36 percent compared with 2011. In 2013, however, domestic prices were raised twice, and as of May 2013 the ratio of domestic to international prices was 1.55 for gasoline and 1.11 for diesel.

Mitigating Measures

- *Fuel subsidies.* Subsidies for the supply of fuels to thermal power plants in Amazonia, a politically sensitive region, were maintained for a period of 10 years, until 2012.

- *Import tax.* In 2001, the government introduced a new tax on the importation and marketing of petroleum products. The levy raised revenues that were then used to fund (1) subsidies for ethanol producers and the transportation costs of hydrocarbons; (2) LPG used by low-income families; (3) projects oriented to environmental protection; and (4) the construction of roads.

- *Gas voucher.* After the withdrawal of LPG subsidies in 2001, the government introduced a new LPG subsidy in 2002 to assist low-income families' purchase of LPG through a gas voucher. Eligibility was based on a means test.

- *Conditional cash transfers.* A conditional cash transfers program, the Bolsa Escola, was implemented in 2001.

Both of these targeted programs (the gas voucher and Bolsa Escola) were consolidated under a new national flagship conditional cash transfer program, the Bolsa Familia, in 2003.

Lessons

A gradual approach in implementing subsidies removal can help minimize the resistance of opposition groups that benefit from subsidies. The phased removal of subsidies in Brazil was carefully tailored to ensure that the process would be politically acceptable. The first products to lose subsidies (asphalt, lubricants, and gasoline for airplanes) were those that generally benefited politically weak stakeholders, and the politically more difficult subsidies (for liquid fuels used for transport and by industry) were removed last.

Liberalization reforms have more chance to succeed with a popular government. After controlling hyperinflation, which had been chronic for over a decade, President Cardoso's administration was able to capitalize on this political support to undertake his liberalization agenda.

Discretionary policies to adjust oil prices and stabilization funds do not work under unstable macroeconomic conditions and can have adverse consequences for the sector. The oil price stabilization fund had run up an enormous deficit in the 1980s, and the government had to transfer an equivalent of 0.8 percent of the 1995 GDP to Petrobras in the middle of the 1990s to pay for oil fund losses. Moreover, underpricing contributed to low investment in exploration and refining capacity.

Macroeconomic instability can contribute to the emergence of subsidies for products with controlled prices. Diesel subsidies emerged in 1999 in the wake of large currency depreciation and the failure to rapidly adjust fuel prices. The liberalization of prices soon afterward allowed the subsidy reform to remain durable, as prices automatically adjusted with fluctuations in the exchange rate.

Targeted social programs can reduce opposition to subsidy reform and enhance its durability. Brazil adopted a gas voucher to compensate low-income households for the increase in LPG prices after the liberalization in 2001, and subsequently it has adopted a conditional cash transfer program, which supports the durability of the subsidy removal.

The incomplete pass-through of fuel prices to consumers can adversely affect the earnings of state-owned enterprises in the energy sector. The profits of Petrobras declined sharply between 2010 and 2012 as the exchange rate depreciated. Even though energy taxes were reduced, this did not fully compensate for rising import prices.

Chile

Context

Chile depends heavily on fossil fuel imports. The share of crude petroleum production to imports has been declining steadily over the past three decades, from 27 percent in 1990 to under 3 percent in 2011.[1] This reflects the combination of shrinking domestic production (which declined by 75 percent in the past two decades) and buoying consumption (which increased by more than 160 percent since the early 1990s), reflecting strong economic growth.

Oil markets in Chile have a long history of deregulation. From the 1920s until the 1970s, the state played a dominant role in Chilean oil markets—from direct involvement in exploration and production to the creation of the national oil company (ENAP). Government involvement kept prices relatively low over this period through implicit subsidies (O'Ryan and others, 2003). In the 1970s, as part of the general push for economic liberalization in Chile, fuel markets (including LGP) were deregulated. This included opening up markets for pro-

TABLE 8.2

Chile: Key Macroeconomic Indicators, 2000–2011

	2000	2003	2008	2010	2011
Nominal GDP per capita (US$)	5174.3	4834.8	10710.7	12570.7	14403.1
Real GDP growth (percent)	4.5	3.4	3.0	6.1	5.9
Inflation (percent)	3.8	2.8	8.7	1.4	3.3
Overall fiscal balance (percent GDP)	−0.7	−0.4	4.1	−0.4	1.3
Central government gross debt (percent GDP)	13.3	12.6	4.9	8.6	11.3
Current account balance (percent GDP)	−1.1	−1.1	−3.2	1.5	−1.3
Oil imports (percent GDP)	2.5	2.8	4.0	2.0	2.6
Oil exports (percent GDP)	0.0	0.0	0.0	0.0	0.0
Oil consumption per capita (liters)	577.3	541.7	833.2	984.2	n.a.
Poverty headcount ratio at US$1.25 per day (PPP) (percent of population)	2.3	2.0	n.a.	n.a.	n.a.
Fuel subsidies (percent GDP)	0.0	0.0	0.0	0.0	0.0

Sources: IEA; IMF, WEO; World Bank, *World Development Indicators*.

---

[1]Balance Nacional de Energia 1988–2011, available at http://bit.ly/GNmVHP.

duction, import, distribution, and sale of fuel products. Nevertheless, ENAP maintains exclusive rights to explore and refine and remains an important player in the oil market. In 2010, ENAP supplied about 70 percent of the Chilean demand for gasoline, diesel, and kerosene.[2]

Reforms Since the Early 1990s

Recognizing the need to smooth the impact of international oil price shocks on domestic consumers, Chile introduced a stabilization mechanism in the early 1990s. Following the spike in oil prices associated with the Gulf War (1990–91), Chile established the Oil Prices Stabilization Fund, Fondo de Estabilización de Precios del Petróleo (FEPP), with an initial fund of US$200 million (0.5 percent of 1991 GDP). Under this mechanism, the authorities set a reference price based on the expected evolution of c.i.f. prices of crude oil in the medium and long term. The fund operated when international prices deviated by more than 12.5 percent from the reference price, by fully subsidizing the difference between international prices and the upper band and imposing a 60 percent tax on deviations below the lower band. The reference price was updated on an ad hoc basis, and the formula behind its calculation was not made public. There was only one fund covering different products (gas, kerosene, diesel, and LPG), which allowed for cross-product subsidization.

The FEPP operated satisfactorily for nearly a decade but required some reforms in the early 2000s to remain financially sustainable (Figure 8.3). The fund remained relatively healthy for the first eight years of operation. However, the sharp increase in oil prices in the late 1990s nearly depleted the fund (the balance reached US$50 million in January 2000), and the mechanism failed to operate in late 1999 (Márquez, 2000). At this point, the fund required an emergency injection of capital to continue operating.

The adjustment mechanism was also modified in a number of ways to increase the financial viability of the fund. These included establishing weekly updates to the reference price (which continued to be based on the current and expected evolution of oil prices in the medium term); introducing an explicit limitation to operate the fund subject to the availability of funds; eliminating the asymmetry in the adjustment mechanism (increasing the tax on deviations below the lower band to 100 percent); increasing transparency by making public the formula to adjust the reference price; and introducing separate funds for gas, kerosene, diesel, and LPG. Nevertheless, even after these adjustments, the fund was nearly depleted by 2003. The total fiscal cost of the FEPP over 2000–2005 is estimated at 0.15 percent of 2012 GDP (Vagliasindi, 2013).[3]

A temporary stabilization fund was established in 2005 in response to supply disruptions. Chile introduced the Fuel Prices Stabilization Fund, Fondo de Estabilización de Precios de los Combustibles (FEPC) as a temporary measure to respond to the spike of prices resulting from the disruption of supply following

---

[2]See http://bit.ly/TsxzGV.
[3]See http://bit.ly/VN9Jo5.

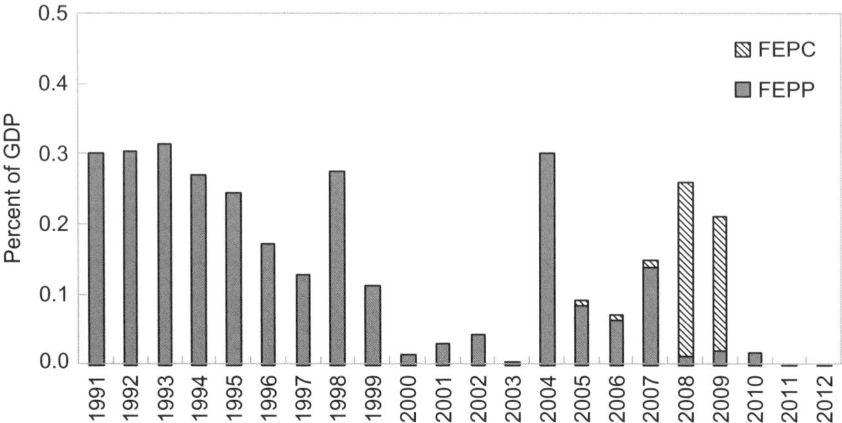

Source: General Treasury of the Republic of Chile (http://bit.ly/Wm0e1e).
Note: FEPC = Fondo de Estabilización de Precios de los Combustibles; FEPP = Fondo de Estabilización de Precios del Petróleo.

Figure 8.3   Chile: Balance of Fuel Stabilization Funds, 1991–2012 (Percent of GDP)

Hurricane Katrina. The mechanism operated in a similar way to the FEPP but relied on a narrower band (5 percent) around a reference price based on the recent and expected evolution of West Texas Intermediate (WTI) prices in the medium term plus a refining fee instead of the price of each derivative product (Organization for Economic Cooperation and Development, 2013). This mechanism was originally intended to be used for about a year but was extended until 2010. The total fiscal cost of the FEPC over 2006–9 is estimated at 0.65 percent of 2012 GDP (Vagliasindi, 2013).

More recently, the stabilization fund was replaced by a tax adjustment mechanism. In 2011, Chile introduced the Consumer's Protection System for Fuel Excise Taxes (Sistema de Protección al Contribuyente ante las Variaciones en los Precios Internacionales de los Combustibles [SIPCO]). Instead of a fund, this adjustment mechanism relies on excise taxes to smooth transmission of changes of international prices to domestic prices. The mechanism reduces excise taxes for fuel when international prices jump above a 10 percent band around a reference price and increasing excise taxes when international prices fall below the band.[4] The reference price is based on the recent and expected evolution of WTI prices in the medium term plus a refining fee for each derivative product. What is important is that by focusing on excise taxes, this excludes large industries (mining, electric generators) that can recover these taxes through deductions (Larrain, 2010).

---

[4]SIPCO was originally introduced with a 12.5 percent band, which was narrowed to 10 percent in September 2012. See http://bit.ly/VRAadr.

Mitigating Measures

Chile has a range of well-targeted safety net programs that it uses to protect low-income groups from economic and other shocks (World Bank, 2010b). In 2005, Chile compensated five million low-income households to offset the impacts of rising fuel prices and another 1.6 million households whose electricity consumption was less than 150 kWh per month. A further payment to low-income families was made in 2006.

Lessons

The costs of smoothing mechanisms depend on their design. For example, there is some evidence that narrowing the bands from 12.5 percent over 1991–2005 to 5 percent over 2006–10 greatly increased cost. In addition, the asymmetric nature of the original adjustment mechanism contributed to the depletion of the fund. These suggest that, when thinking about the parameters of an adjustment mechanism, specific details can have a great impact on the cost of these programs. Thus, countries considering introducing these smoothing devices should carry out illustrative scenarios, including sensitivity analysis of the parameters, to ensure that the cost of the program would be in line with expectations.

Adjustment mechanisms should be transparent. Initially, the FEPP used a secret formula and allowed for ad hoc adjustments in the reference band. This added unnecessary uncertainty regarding the timing and size of future fuel price adjustments and the extent to which international shocks would be transmitted to local prices. Such uncertainty is at odds with the goal of stabilizing prices. The reform of the early 2000s shows that it is possible to use a transparent rules-based approach to meet these goals.

It is possible to target the smoothing adjustment to smaller consumers. One important characteristic of the latest Chilean reform is that it excludes large energy consumers by applying the adjustment through an excise tax that is generally deducted by industries in mining, electricity, and other large fuel consumers. This sends a clear signal that these large consumers should be able to hedge on their own and helps to buy in support for reforms from the general population.

Smoothing mechanisms should offer only temporary relief. In Chile, the fuel market has been liberalized since the 1970s. Thus, these smoothing mechanisms have been in part a product of popular outcry related to higher fuel prices (in the context of the Gulf War and Hurricane Katrina, for example). Nevertheless, Chile has used these mechanisms for temporary support—all of the adjustment schemes intended that the increases to international prices be eventually transmitted to local prices in full. It is important to note that Chile has achieved this while devoting important resources to a well-targeted safety net (World Bank, 2010b).

Peru

Context

Peru is a net importer of oil, and its import bill is highly dependent on developments in international prices. Diesel accounts for the largest share of fuel consumption (47 percent) followed by LPG (19 percent). Historically, the consumer prices of these commodities have been politically sensitive. Diesel is primarily consumed by public transportation vehicles, and most households use stoves fueled with LPG.

Peru's oil market is a duopoly controlled by two companies responsible for refining and distribution, the private Relapasa and the public PetroPerú. Before the creation of the fuel product stabilization fund in 2004, the authorities regulated consumer prices by managing the prices of fuels sold by PetroPerú. Because PetroPerú controlled a significant share of the market, Relapasa had to adjust its prices based on that benchmark, incurring losses when international prices were above domestically regulated retail product prices.

Reforms Since 2004

Price smoothing mechanism. In May 2004, beleaguered by increasing prices amid a global hike in commodity prices, a smoothing price mechanism was put in place. The mechanism sought to smooth changes in domestic prices by adjusting excise taxes. Excise taxes were adjusted downward (upward) each time international prices breeched an upper (lower) price band in order to keep the consumer price constant. However, this initial attempt to limit the pass-through from international to domestic prices did not perform well, mainly because rising prices created sizeable revenue losses. These losses, in turn, caused liquidity shortfalls in the treasury because of the shortfall in expected revenues.

In September 2004, the authorities created a stabilization fund—the Fondo de Estabilización de Precios de Combustibles (FEPC). The fund involved a complex payment system financed directly by the treasury. All types of gasoline and LPG were included as products whose prices were to be regulated by the FEPC. The objective of the FEPC was to prevent the full transmission of international to domestic prices. This was to be achieved by providing transfers to the refineries in periods when international prices were rising to compensate them for their increase in supply costs. When reference prices were above the upper limit of the

TABLE 8.3

Peru: Key Macroeconomic Indicators, 2000–2011					
	2000	**2003**	**2008**	**2010**	**2011**
Nominal GDP per capita (US$)	1334.7	2258.9	4481.5	5290.8	6007.9
Real GDP growth (percent)	2.8	4.0	9.8	8.8	6.9
Inflation (percent)	3.76	2.3	5.8	1.5	3.4
NFPS balance (percent of GDP)	−3.4	−1.7	2.4	−0.3	1.9
Gross public debt (percent of GDP)	n.a.	46.9	25.9	23.3	21.2
Current account (percent of GDP)	−2.9	−1.5	−4.2	−2.5	−1.9
Oil imports (percent of GDP)	3.1	2.3	4.1	2.6	3.2
Oil exports (percent of GDP)	1.1	1.0	2.1	2.0	2.7
Oil consumption per capita (liters)	n.a.	332.5	380.2	377.2	317.8
Poverty headcount ratio at US$1.25 per day (PPP) (percent of population)	n.a.	9.5	6.2	4.9	n.a.

Sources: IEA; IMF, WEO; World Bank, *World Development Indicators.*
Note: NFPS = Non-financial public sector.

price band, a contingent credit for refineries (payable by the treasury) was created. Similarly, if reference prices were below the lower limit, a contingent liability for refineries (payable to the treasury) was created.

The performance of the FEPC prior to the reforms was mixed. Although it has been successful in limiting the pass-through from international to domestic prices, the FEPC has also generated sizeable fiscal costs (Figure 8.4). The latter is the consequence of the upward trend in oil prices and the authorities' reluctance to increase the upward limit of the band. This combination of higher prices and frozen bands has proven to be a drain on fiscal resources. Another problem has been the accumulation of contingent liabilities. There is no legal obligation of the treasury to pay the refineries; instead, the treasury has paid when it has had sufficient cash on hand. This has created acute liquidity issues, particularly in 2008, for the refineries, which have made repeated requests to the treasury to honor its obligations.

Stabilization fund. Reforming the FEPC had long been an objective of the authorities. By mid-2008, when the FEPC had accumulated a record amount of liabilities (equivalent to the total cost of the extreme-poverty alleviation program), the authorities disseminated a study on the distributional impact of the subsidy. The study confirmed the regressive impact of this untargeted subsidy. It found that the subsidy received by the wealthiest 20 percent of the population was eight times the amount received by the poorest. The study received widespread media coverage. Nevertheless, the authorities could not reach a consensus among stakeholders to proceed with a comprehensive reform, although they managed to implement a modest increase in the pricing bands.

In 2010, amid a reduction in international prices, the authorities saw a window of opportunity to introduce reform measures. In April they introduced a rule to

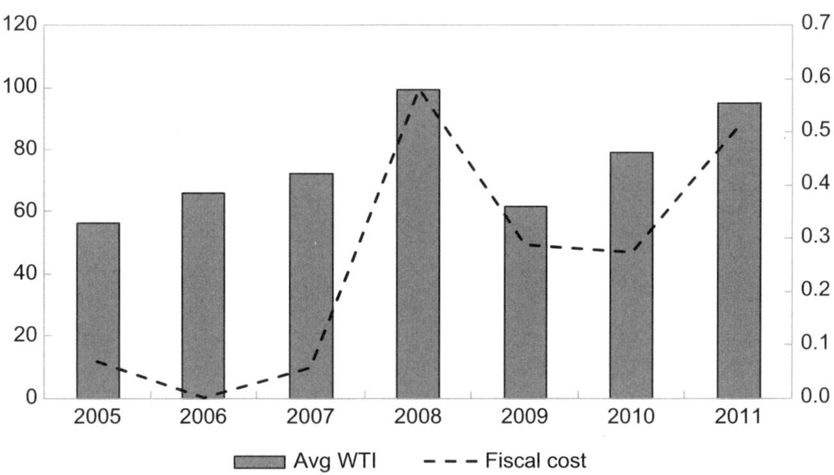

Source: IMF staff estimates based on data from the Peruvian authorities.
Note: Avg WTI = average based on the West Texas Intermediate.

Figure 8.4 Peru: International Price and the Fiscal Cost of Fuel Subsidies, 2005–11

TABLE 8.4

Peru: Spending of the Oil Price Stabilization Fund by Type of Product, 2011

	Million US$	Percent
Total	**871.8**	**100.0**
Diesel	440.6	50.5
LGP	261.0	29.9
Gasolines	106.5	12.2
Industrial petroleum	28.1	3.2
Gasohols	35.7	4.1

Source: Country authorities.

automatically update the band limits every two months. Price changes would nevertheless be limited to 5 percent, with an exception for domestic consumption of LPG, whose maximum price change was 1.5 percent. The authorities also created a special subaccount in the treasury to finance the FEPC, thus reducing uncertainty regarding payments to refineries. In October 2011, all types of high-octane gasoline, which is used by luxury cars, were excluded from the FEPC, with international price changes being fully passed on to domestic prices. In August 2012, regular gasoline was also removed, with only diesel and LPG for household consumption remaining (LPG for industrial consumption was excluded).

The reform has been successful in reducing the fiscal cost of the subsidy without provoking widespread opposition. At the same time, the reform did not touch on the most politically sensitive products, diesel and LPG, which also represent the largest share of subsidy spending (80 percent—see Table 8.4). As a result, the total fiscal savings of the reform have been modest: around 0.1 percent of GDP.

Mitigating Measures

Mitigating measures were not implemented, because reforms did not reduce subsidies for the products most heavily consumed by the poor.

Lessons

Regulating consumer prices by adjusting taxes can lead to challenges for fiscal management. Fuel products are widely consumed, so the fiscal impact of even minor changes in taxation could be significant. Although a smoothing mechanism can help shield households from the shock of higher oil prices, it can create challenges for fiscal management, even when there is fiscal space to accommodate some price smoothing. For example, the transfer of resources from the treasury to the FEPC put pressure on the treasury's liquidity and complicated its cash management, particularly because of the strong seasonal behavior of revenues and spending in Peru. Varying commodity taxation according to international prices can also create uncertainties with respect to projected revenues, given the volatility of commodity prices. To address these concerns, either more automatic adjustment mechanisms or a larger reserve of funding built up during good times is needed.

Price smoothing mechanisms should incorporate automatic adjustments of pricing bands. The core principle of a price smoothing mechanism is that it moderates only volatility. However, if prices are on an upward trend, the smoothing mechanism must have some method to adjust to this. In the case of Peru, the decision to keep the bands untouched in the wake of an upward trend in oil prices has proven to be fiscally costly. This has in effect converted a price smoothing mechanism into a pure subsidy.

Rules for the payment of refineries for subsidies should be explicit. This can be done with a special subaccount, which ideally should be integrated into the Treasury Single Account to ensure transparency. This will make explicit the size of the subsidy and also give certainty to the amounts that could be compensated to refineries.

Introducing subsidy reforms during "good times" can enhance the chances of success. The decision to introduce the reform in early 2010, during a period of stable prices and strong economic growth, helped make the reform politically more palatable.

To ensure public support, subsidy reforms can most fruitfully begin with the products consumed most heavily by higher-income groups. In the case of Peru, this meant starting the reform by raising high-octane gasoline prices. Despite the fact that fiscal savings under this approach can be limited, such a strategy may be warranted to allow stakeholders to see the effects of the reform and allow more time to muster support for a broader reform. This approach also signals the direction of reform and can pave the road for further, more ambitious reforms. There is nevertheless a trade-off involved between fiscal savings and safeguarding against adverse effects on lower-income groups, as indicated by the modest savings that subsidy reform has achieved thus far for Peru.

ELECTRICITY SUBSIDIES

Brazil

Context

During the 1980s, Brazil's economy was characterized by low growth and macroeconomic imbalances. Growth averaged about 3 percent and inflation was high, averaging 272 percent. Fiscal policy was expansionary, with the overall budget deficit averaging 5 percent of GDP during the period and reaching 7 percent in 1989. The weak fiscal performance led to an increase in net public debt from 24 percent of GDP in 1981 to almost 40 percent in 1989. These deteriorating conditions put pressure on the authorities to alter Brazil's import-substitution policies and liberalize the economy (Giambiagi and Moreira, 1999).

Privatizing the power sector was part of the authorities' reform efforts in this area and was attractive for three reasons. First, privatizing some of the assets of the sector held out the promise of raising substantial revenues for the treasury and clearing debt from the federal government off the books. In 1993, this external debt contracted by the electricity companies amounted to almost 25 percent of Brazil's external debt. Second, selling off the state-owned distribution companies meant that a significant amount of state-level debt owed to the national government would be paid. Third, the federal government believed that it would be difficult to raise sufficient amounts of capital on its own to invest in the facilities needed to meet growing demand. Electricity investment dropped by almost half, in the 1990s, relative to the decade before.

Other sector-specific factors also motivate the privatization. These included high construction costs incurred because of cartels among contractors, excessive employment, and high power losses throughout the system. Finally, private distribution companies would establish tariff structures that were more reflective of costs.

The performance of the sector in the 1980s was poor and provided the primary impetus for structural change. A number of factors contributed to this weak performance, including the regulatory framework for determining prices. This was determined by the Planning Secretary of the President's Office. Price adjustments were influenced by the desire to contain inflation and were unrelated to developments in costs and the need to ensure an adequate rate of return to capital. This approach to pricing resulted in declining electricity tariffs in real terms, and it undermined incentives for improving productivity. It led to a worsening of the companies' financial positions and an increase in external debt to finance expansion of the electricity supply. As a result, debt accumulated in the CRC (Conta de Resultados a Compensar) reached US$26 billion in 1993, which was absorbed by the central government in the same year. Assuming this represented the accumulation of losses over the preceding five years, the subsidy to the electricity sector had averaged 0.7 percent of GDP per year in 1987 (Figure 8.5).

Reforms Since 1993

The overall plan devised for the power sector was for all assets to be privatized to the fullest extent possible. In order to remove constraints for privatization, in 1993 the requirements of uniform national tariffs and of a mandated 10 percent rate of return on capital were removed. Significant pricing reforms were also undertaken. To increase the transparency of the pricing system in the electricity sector, a law was enacted in 1998 to unbundle the electricity sector system.

The government decided to begin the electricity sector privatization with the distribution companies. This was motivated by the fact that substantial productivity gains could be realized in these activities. In addition, the financial problems in these companies had ripple effects in the entire sector, because financially insolvent distribution companies were not paying electricity generation companies. Fixing the finances of the distribution companies and making them creditworthy buyers of energy had a positive effect on the power generation sector and helped pave the way for privatizing these assets.

The privatization program took place over a 10-year period, from 1993 to 2003, and resulted in a competitive generation market with a number of private companies competing. The distribution sector was privatized under a series of monopoly licenses, although over time the end users could obtain third-party access to the grid and the industry was under regulatory jurisdiction.

The 1993 reforms were successful, from a fiscal point of view, in eliminating subsidies. However, the privatization of the sector was not accompanied by a

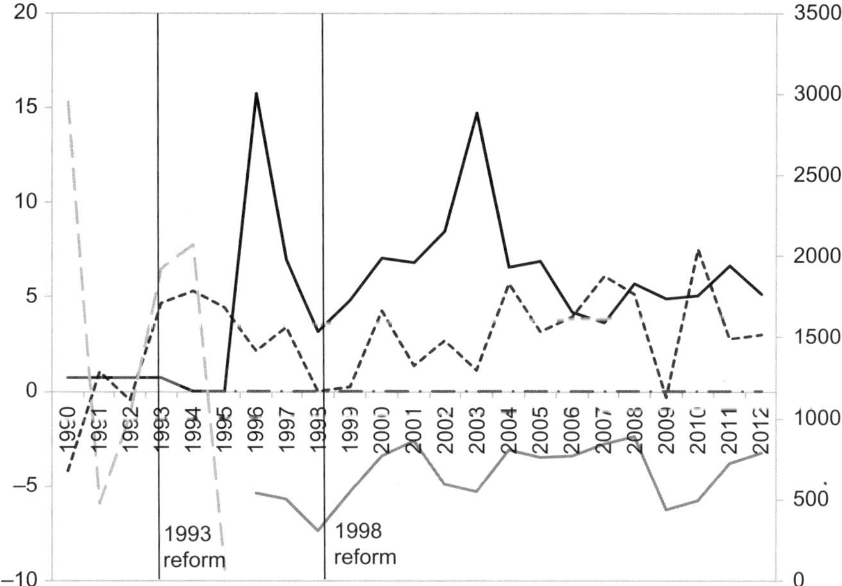

Sources: IMF staff estimates; IMF WEO database.
Note: Data on fuel subsidy were calculated from the average of debt accumulated under the Fuel Consumption Account from 1987 to 1993 and paid by the treasury in 1993; from 1990 to 1994 annual inflation was over 500 percent.

Figure 8.5   Brazil: Macroeconomic Developments and Electricity Subsidy Reforms, 1990–2012 (Percent of GDP or rate)

strong regulatory framework. This led to an uncertain investment climate and the suspension in the construction of some distribution lines. The lack of investment in electricity generation, combined with a drought in 2001, caused Brazil's hydro-electricity reservoirs to become dangerously depleted. To avoid a larger energy supply crisis, the government introduced regulations that forced producers to ration the electricity they supplied to consumers and allowed distributors to raise their tariff levels to compensate for their losses during the rationing period. These decisions produced a sudden drop in GDP and a steep increase in tariff levels. This undermined public support for privatization and contributed to a slowdown in progress in liberalizing the energy sector.

Even after the privatization, cross-subsidies were maintained. They were designed to support rural electrification and establish special rates for low-income households. But no uniform method to implement these cross-subsidies was developed, and each concessionaire was at liberty to fashion its own. The result is a potpourri of subsidies whose targeting efficiency is almost impossible to measure.

In summary, the electricity market reform in Brazil was successful in many respects. It eliminated government subsidies to the sector, depoliticized tariff increases, secured the expansion of electricity generation (post-2001), and reduced vulnerabilities associated with the external debt acquired by electricity sector companies. Mota (2003) evaluated the effects of the electricity sector privatization on supply quality and cost and found that the efficiency gains resulting from cost reduction were substantial. These were obtained through the reduction, by half, in the number of employees from 1994 to 2000. With respect to the impact of privatization on quality, there has been an improvement in security and availability of energy supply.

Mitigating Measures

- *Cross-subsidies.* Even after the liberalization of the sector, regional cross-subsidies remained.

- *Regional fuel subsidy.* A levy on electricity tariffs was introduced in 1993 to subsidize the supply of fuels to the inefficient thermal power plants of Amazonia, a politically sensitive region, and was maintained for an extended period.

- *Income-based tariff relief.* In 1995, legislation was approved to provide lower electricity tariffs for low-income households.

- *Free power to rural areas.* In 2003, the government introduced a program to finance free electricity to ten million rural people, which is funded by levies on electricity tariffs.

Lessons

The need to correct macroeconomic imbalances can provide political support for reforms. The low economic growth, hyperinflation, and high external debt burden of Brazil in the 1980s forced politicians to react and consider subsidy reform as an option to confront these imbalances.

Controlling increases in electricity prices as an anti-inflationary tool can have adverse fiscal consequences. The adoption of this policy in the 1980s resulted in financial losses for the sector, the accumulation of debt, and underinvestment.

Reforms have a better chance of success with a popular government. After controlling the hyperinflation, which had been chronic for over a decade, President Cardoso's administration was able to capitalize on this political support to undertake his agenda for energy sector liberalization.

Targeted social programs can reduce opposition to subsidy reform and assist the poor. Brazil has adopted a policy to reduce electricity tariffs for low-income people and has adopted a conditional cash transfer program, which facilitated the implementation of the subsidy reforms.

Privatization of electricity companies without a strong regulatory framework can have serious consequences and undermine popular support for energy reform. The unclear rules in the first years of the privatization process resulted in low levels of investment and contributed to the energy crisis in 2001.

TABLE 8.5

Mexico: Key Macroeconomic Indicators, 2000–2011					
	2000	**2003**	**2008**	**2010**	**2011**
Nominal GDP per capita (US$)	6858.8	6864.7	10050.5	9218.5	10153.3
Real GDP growth (percent)	6.0	1.4	1.2	5.5	4.0
Inflation (percent)	9.5	4.6	5.1	4.2	3.4
Overall fiscal balance (percent GDP)	−3.1	−2.3	−1.1	−4.3	−3.4
Public debt (percent GDP)	42.6	45.6	43.1	42.9	43.8
Current account balance (percent GDP)	−2.8	−1.0	−1.4	−0.3	−0.8
Oil imports (percent GDP)	1.1	1.2	3.3	2.9	3.7
Oil exports (percent GDP)	2.4	2.7	4.6	4.0	4.9
Oil consumption per capita (liters)	505.6	529.9	653.8	607.0	n.a.

Sources: IEA; IMF, WEO; World Bank, *World Development Indicators*.

Mexico

Context

Mexico has a sound macroeconomic policy framework but suffers from high income inequality and poverty. Fiscal and monetary fiscal policies are underpinned by a fiscal rule and inflation targeting. Mexico's Gini coefficient averaged 0.48 in the late 2000s, which indicates a significantly higher degree of inequality than the OECD average. About 46 percent of Mexico's total population lives in poverty, while about 10 percent lives in extreme poverty.

The electricity sector is dominated by the government-owned Comisión Federal de Electricidad (CFE). CFE is a major electricity generator, accounting for about three-quarters of total generation capacity of the country, and monopolizes transmission and distribution functions.[5] Although independent power producers entered the market after deregulation in the generation sector in 1992, they account for only about one-quarter of generation assets. The dominance of the public sector in the electricity market is mandated by constitutional provisions.[6] The Comisión Reguladora de la Energia is the regulator of the electricity sector.

Electricity tariffs have been set below cost-recovery levels. A study of tariff structures in 2005–6 showed that tariffs were below cost-recovery levels for most residential users (by about 40 percent) and the agricultural sector (by about 30 percent). Subsidies were smaller for other sectors, but tariffs still failed to cover costs. The benefit incidence of these subsidies is highly regressive (Komives and others, 2009). The Ministry of Finance and Public Credit has tariff-setting authority, and tariffs have been adjusted monthly in propor-

---

[5]Before 2009, electricity distribution was operated under a duopoly by CFE and Luz y Fuerza del Centro (LFC), a government-owned company that served customers in metropolitan Mexico City. In 2009, the government closed down LFC to eliminate direct government subsidies to LFC to cover operating losses and let CFE take over LFC's service areas.

[6]The Public Electricity Service Act, which was amended in 1992 and opened electricity to the private sector, lists the following areas as falling outside of "public service" and thus open to private sector participation: self-supply, cogeneration, independent power production, imports, exports, and small-scale generation.

tion to changes in input prices for electricity generation, transmission, and distribution, rather than based on actual service costs. Tariff systems are highly complex, with over a hundred different billing possibilities for residential users. They are built on block tariffs, which provide larger subsidies for users who consume less. In addition, a scheme of "summer subsidies" provides a discount to residential customers residing in warm areas to compensate for higher air-conditioning costs.

Electricity subsidies impose a substantial fiscal burden. Electricity subsidies were estimated at about 0.5 percent of GDP in 2011, a similar ratio as 10 years earlier (Figure 8.6).[7] The government does not record the subsidies explicitly. Under the so-called *aprovechamiento* system, CFE must pay the government a return on the fixed assets (9 percent), but this is transferred back from the government to CFE to cover tariff subsidies and infrastructure investment (OECD, 2004). Since 2002, the amount collected under such a system has fallen short of what was needed to cover tariff subsidies, thus eroding CFE's capital base (Komives and others, 2009).

Reforms Since 1999

Reform initiatives for the electricity sector and subsidies have been unsuccessful.

A comprehensive reform proposed in 1999, to include market privatization, failed on account of legal impediments, opposition from interest groups, lack of public awareness, and political impasse. In 1999, President Zedillo proposed a comprehensive reform package that included unbundling of generation, transmission, and distribution; creation of a wholesale market; privatization; and strengthening of the regulator's power. It failed for a number of reasons. These included legal impediments, such as the need for a constitutional amendment to allow broad private sector participation; opposition from powerful interest groups, mainly consumers and labor unions for CFE employees, who opposed tariff reform and privatization; limited public awareness about problems in the electricity sector and public opinion against privatization;[8] and a political impasse in the period leading up to the 2000 presidential election (Carreón-Rodriguez, San Vicente, and Rosellón, 2003).

A reform proposal launched by President Fox in April 2001, although it de-emphasized privatization, also failed. The president could not forge a consensus in the congress to turn the bill into a law. In addition to the obstacles against President Zedillo's reform, President Fox also had to cope with political fragmentation after a drastic change in the political landscape. In particular, after

---

[7]This estimate of the cost of subsidies, from the authorities, exceeds the figure provided by the IEA, which indicates subsidies were about 0.1 percent of GDP or smaller in 2007–10. The reason is that the IEA approach only measures consumer subsidies using the price-gap approach and does not measure the total budget support that also compensates producers for inefficiencies (producer subsidies).

[8]According to a public opinion poll in 2002, 49 percent of respondents acknowledged problems in the electricity sector. Thirty-five percent of respondents opposed private investment, while 17 percent supported a strategy of encouraging it (Carreón-Rodriguez, San Vicente, and Rosellón, 2003).

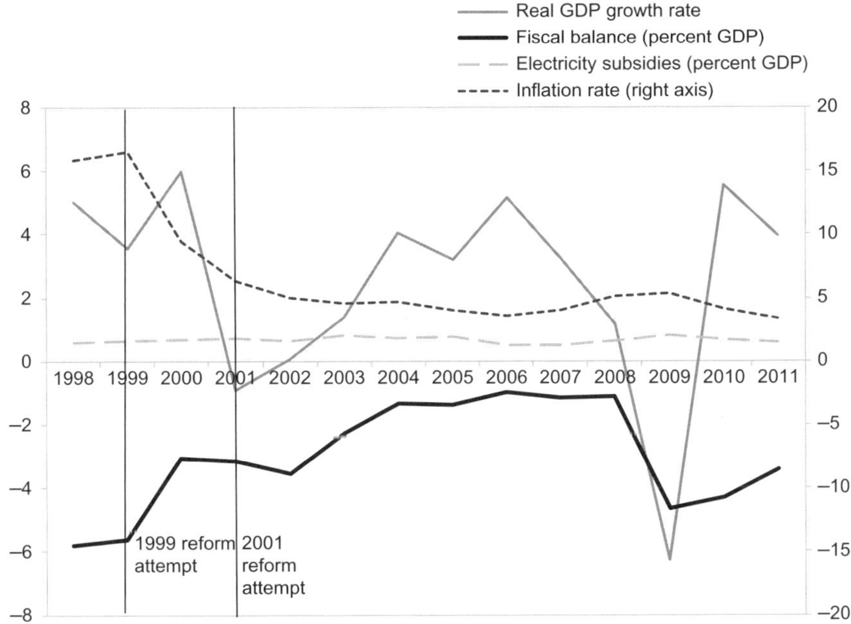

Sources: Comisión Federal de Electricidad; IMF staff estimates; IMF, WEO.

Figure 8.6 Mexico: Macroeconomic Developments and Electricity Subsidy Reforms, 1998–2011

seven decades of continuous ruling, the Partido Revolucionario Institucional (PRI) was defeated in the 2000 presidential election, and political parties were forced to compromise with labor unions and powerful conglomerates, which had earlier been submissive to presidential administrations under the PRI. Various reform proposals made by opposition parties, as well as their petition to the Supreme Court on the constitutionality of President Fox's proposal, complicated the debate.

Tariff reform was implemented in 2002 to reduce electricity subsidies. In particular, a tariff was introduced that exceeded the long-term marginal cost for customers with high consumption volumes. However, the reform did not lead to a permanent reduction in subsidies, as the scheme of "summer subsidies" allowed customers to be reclassified into highly subsidized categories.

Mitigating Measures

- *Tariff structure.* The residential tariff structure is characterized by an extensive list of subsidized categories. Tariffs are subsidized for customers who consume less and reside in warm areas. The latter, called a scheme of "summer subsidies" applicable during the summer season, classifies customers into six categories based on average real temperatures (cutoffs are 25°C, 28°C,

30°C, 31°C, 32°C, and 33°C), with customers living in warmer areas receiving higher subsidies. More customers were reclassified from lower-temperature categories to higher-temperature ones during the 2000s, further increasing the overall subsidy bill (Komives and others, 2009).

- *Social safety net, including cash transfers.* Mexico has a well-developed safety net program, Oportunidades, that has not yet been used in the context of subsidy reform. Oportunidades is a cash transfer targeted for families of extreme poverty and is conditional on school attendance and medical checkups of family members. In 2008, about five million families benefited from the program. Benefits consist of not only direct cash transfers and education grants but also cash compensation for energy consumption expenses. Oportunidades offers more effective and better-targeted pro-poor subsidies than fuel and electricity subsidies do (authorities also acknowledge that the incidence of the electricity subsidy is highly regressive), while it cost only one-fifth of total fuel subsidies (including subsidies for petroleum products and electricity) in 2008.

Lessons

The failure of the electricity sector reform in Mexico reveals the numerous obstacles to successful reform. A long history of tariff subsidies and the vertical and horizontal dominance of a state-owned company created strong interest groups against reform, especially consumers and labor unions. Political fragmentation, the constitutional mandate for the public sector to run the electricity sector, and public opinion against privatization made the reform even more challenging. The extensive list of subsidized customer categories has contributed to an increase in overall subsidies, as customers have been reclassified into highly subsidized categories. Mexico's case also suggests that the existence of a targeted safety net and commitment to sound macroeconomic policies are not sufficient to successfully reform electricity subsidies. A thorough public information campaign, as well as transparent accounting of electricity subsidies, would be a key first step for successful reform.

CHAPTER 9

Case Studies from Central and Eastern Europe and the Commonwealth of Independent States

Katja Funke, Kangni Kpodar, and Baoping Shang

PETROLEUM PRODUCT SUBSIDIES

Turkey

Context

Prior to reforms, the Turkish petroleum sector was dominated by state-owned vertically integrated enterprises. Before 1990, the public distribution company Petrol Ofisi and the public refining company TÜPRAŞ were subsidiaries of Türkiye Petrolleri Anonim Ortaklığı (TPAO), the public petroleum exploration and production company. At that time, the industry was governed by public decrees under which prices of petroleum products were set largely by the government.

The petroleum sector reform started in the 1980s as part of broader economy-wide reforms moving toward a market-oriented regime. The policy regime prior to these reforms involved heavy state intervention in economic activities, in particular in the form of government ownership of enterprises in critical industries, such as energy, telecommunications, petrochemicals, iron, and steel. The state also played a critical role in the allocation of financial resources, especially through state-owned banks. However, after a major balance of payments crisis in the second half of the 1970s and a military coup in 1980, Turkey was determined to transform its economy into a more market-oriented regime, through mass liberalization of domestic markets and international trade.

Reforms Since 1989

The petroleum sector reform aimed to achieve several objectives:

- *Improve the fiscal position of the government.* The reform would eventually eliminate petroleum subsidies, both consumer and producer subsidies.

- *Reduce the inefficiencies in the petroleum sector.* Private participation would introduce competition, improve efficiency, and limit monopoly abuse in the sector.

TABLE 9.1

Turkey: Key Macroeconomic Indicators, 2000–2011

	2000	2003	2008	2010	2011
Nominal GDP per capita (US$)	4146.8	4534.9	10272.4	10062.4	10521.8
Real GDP growth (percent)	6.8	5.3	0.7	9.0	8.5
Inflation (percent)	55.0	25.3	10.4	8.6	6.5
Overall fiscal balance (percent GDP)	n.a.	−10.0	−2.4	−2.7	−0.3
Public debt (percent GDP)	51.6	67.7	40.0	42.2	39.4
Current account balance (percent GDP)	−3.7	−2.5	−5.7	−6.3	−9.9
Oil imports (percent GDP)	3.6	3.8	6.6	5.2	7.0
Oil exports (percent GDP)	0.1	0.3	1.0	0.6	0.6
Oil consumption per capita (liters)	254.7	246.0	310.1	304.6	n.a.
Poverty headcount ratio at US$1.25 per day (PPP) (percent of population)	n.a.	2.5	0.0	n.a.	n.a.
Fuel subsidies (percent GDP)	0.0	0.0	0.0	0.0	0.0

Sources: International Energy Agency; IMF, *World Economic Outlook* (WEO); World Bank, *World Development Indicators.*
Note: n.a. = not applicable; PPP = purchasing power parity.

- *Meet the preconditions for Turkey's European Union (EU) membership.* The reform was also urged by various international institutions that provided support during several economic crises.

Turkey initiated a series of reforms that can be characterized as a long process toward full price liberalization, privatization of state-owned enterprises, and a competitive energy market.

Under a new law passed in 1989, private companies were allowed to set prices, and in 1990 public companies began to be privatized. Under the 1989 law, importers, refining companies, distribution companies, and retailers were, in theory, to be allowed to set the prices of crude oil and petroleum products. The privatization process of public refining and distribution companies started in 1990 and was fully completed in 2005. This did not, however, achieve a liberalization of prices at the time. The reason was that the government maintained control of the state-owned enterprises that dominated the petroleum product market, which in practice set the prices of petroleum products—even though a liberal price regime was adopted legally. These reforms were adopted when the government was led by the Motherland Party, a center-right nationalist party that supported restrictions on the role government could play in the economy and favored private capital and enterprise.

In 1998, the automatic pricing mechanism was adopted by the government, which set a ceiling on the prices of almost all oil products based on international petroleum prices and the exchange rate. In principle, refining companies and importers could set prices freely, provided these prices did not exceed the ceilings. However, there were still license requirements for importing and capacity requirements for storage, and these requirements presented large barriers for market entry. In practice, distribution companies and retailers were not allowed to set their prices freely; instead, prices were set by the government TÜPRAŞ, the public refining company, benefited significantly from the automatic pricing mechanism and was able to make profits. TÜPRAŞ had often incurred losses before

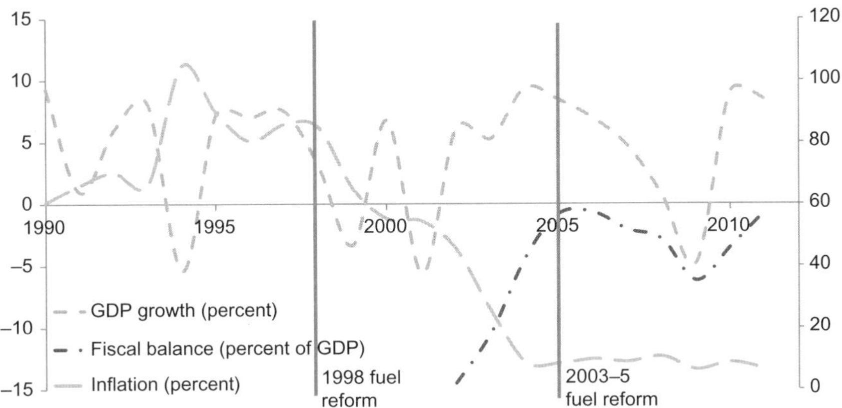

Figure 9.1 Turkey: Macroeconomic Developments and Energy Subsidy Reforms, 1990–2011

the mechanism, because the government kept the prices of petroleum products low. The automatic pricing mechanism reform was also under the watch of the Motherland Party, whose popularity, however, had declined significantly.

In 2003, regulatory authority over the petroleum product market moved to an independent agency. The Petroleum Market Law was passed in 2003 to achieve the institutionalization of the market economy and to comply with EU legislation and other international obligations. The law took the regulatory authority of the petroleum market from the Ministry of Energy and Natural Resources and placed it under the control of the Energy Market Regulatory Authority, an independent agency that was also the regulator of the electricity market at the time. Under the Petroleum Market Law, government control of the petroleum market, such as through license requirements and importation limits, was loosened. The privatization of state-owned enterprises was also accelerated under the law and was completed by 2005.

The most important impact of the Petroleum Market Law was the full liberalization of fuel prices, which came into effect in 2005 (Figure 9.1). Since then, fuel prices have been set by the market. Turkish gasoline and diesel prices are now among the highest in the Organisation for Economic Co-operation and Development, owing to the relatively high excise taxes that are reflected at the level of retail prices (Figure 9.2) (IEA, 2010).

The 2003 and 2005 reforms were introduced when the Justice and Development Party was in power. The Justice and Development Party is a center-right conservative party that came to power in 2002 by a landslide victory and has since maintained a strong majority in the parliament.

Mitigating Measures

In addition to the existing social safety net programs, several targeted measures were taken to mitigate the impacts of reforms:

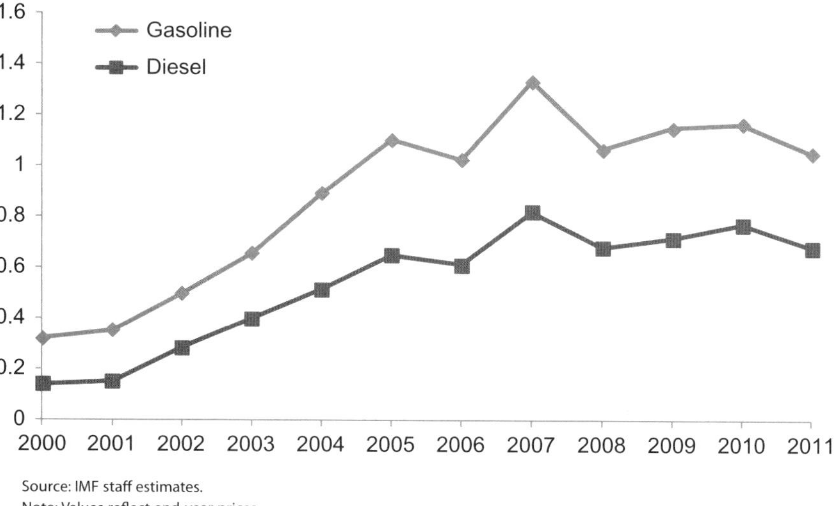

Source: IMF staff estimates.
Note: Values reflect end-year prices.

Figure 9.2 Turkey: Gasoline and Diesel Net Tax, 2000–2011 (U.S. dollars per liter)

- *Tax exemption for liquefied petroleum gas (LPG) consumption.* Between 1999 and 2001, the government supported the use of LPG by households for cooking purposes by forgoing both value added tax (VAT) and the special consumption tax. These tax exemptions resulted in the price of LPG being below that of both gasoline and diesel. As regular motor engines cannot use LPG, the government expected the fuel's use in cars to remain limited. However, an underground industry soon developed to make gasoline and diesel engines compatible with LPG. With a payback period of less than two years, the operation proved sufficiently simple and cheap for drivers to convert their vehicles to LPG use. Alerted by the resulting loss of tax revenue, the government began to phase out this tax expenditure at the end of 2000. This provision resulted in significant increases in LPG consumption.

- *Tax exemption for public transportation.* According to the New Turkish Corporate Tax Law passed in 2006, public transport companies owned and managed by municipalities, villages, or special provincial administrations are exempt from VAT and excise tax.

- *Rebate for diesel used in agriculture.* In Turkey, the tax rate on diesel fuel is very high, thereby affecting the real income of farmers. A rebate program was introduced by the Ministry of Agriculture in 2007 to help farmers grow specific crops. There are three different types of crops defined by the ministry, which correspond to different aid levels. The amounts of aid are calculated according to the area of the land used in growing specified crops and paid according to a schedule defined by the cabinet. There are no restrictions on how grant money is spent. The measure is to be phased out.

Lessons

Broad support for and firm commitment to market reform was key to the success of subsidy reforms in Turkey. Turkey started a more liberalized regime for energy pricing in the late 1980s and early 1990s and sustained these reforms under the administration of various political parties. Aided by economy-wide reforms to enter the EU, energy sector reforms have received broad support with little setback. Because of this, only very limited mitigating measures were adopted and the popularity of government did not appear to have a large bearing on the success of energy reforms.

Improving economic conditions also helped advance reforms. In the past two decades, the Turkish economy has grown steadily; inflation has been lowered substantially; and the overall fiscal balance has also been improving. The short-term impact of energy reforms on household welfare has been limited because of relatively high household income. This assured the public that the country was moving in the right direction and prevented any setback to reforms from occurring.

Independent agencies for energy policy can help steer technical decisions away from politics. Under the Petroleum Market Law, an independent agency, the Energy Market Regulatory Authority, was responsible for implementing the laws and regulating the petroleum sector. This took the technical decisions on pricing and market regulation out of the hands of politicians and ensured the stability and consistency of reforms.

ELECTRICITY SUBSIDIES

Armenia

Context

In the early 1990s, Armenia began the transition to a market-based economy with a financially weak electricity sector. The industry was dominated by a vertically integrated and monopolistic power company and characterized by heavily subsidized

TABLE 9.2

Armenia: Key Macroeconomic Indicators, 2000–2011

	2000	2003	2008	2010	2011
Nominal GDP per capita (US$)	593.5	874.1	3,605.9	2,840.4	3,032.8
Real GDP growth (percent)	5.9	14.1	6.9	2.1	4.4
Inflation (percent)	−0.8	4.7	9.0	7.3	7.7
Overall fiscal balance (percent GDP)	−6.3	−1.5	−1.8	−4.9	−2.7
Public debt (percent GDP)	48.9	32.9	14.6	33.3	35.1
Current account balance (percent GDP)	−14.6	−6.8	−11.8	−14.7	−12.3
Oil imports (percent GDP)	0.1	0.1	0.4	0.3	0.3
Oil exports (percent GDP)	0.0	0.0	0.0	0.0	0.0
Oil consumption per capita (liters)	133.7	139.3	159.5	127.6	n.a.
Poverty headcount ratio at US$1.25 per day (PPP) (percent of population)	n.a.	10.6	1.3	n.a.	n.a.

Sources: IEA; IMF, WEO; World Bank, *World Development Indicators*.

retail prices. The sector was largely dependent on fuel imports from other countries of the former Soviet Union. The collapse of the former Soviet Union and the conflict with neighboring Azerbaijan led to severe disruptions in oil supply. Electric generation declined by almost 50 percent in 1990–95, resulting in chronic power shortages.

Fiscal and quasi-fiscal subsidies were large and unsustainable. After the transition to a market economy, the power generation mix shifted from oil to hydropower. Despite the fact that the latter is a cheaper source of electricity production, electricity subsidies remained very large, amounting to about 11 percent of GDP in 1995. There were various forms of subsidies:

- *Implicit consumer subsidies owing to low retail prices.* Prices were set below levels needed to cover operating costs and capital depreciation. Because there were no transfers from the budget to cover these costs, the electricity companies financed them by accumulating debt with the banking system. Electricity tariffs for households included a cross-subsidy from other energy users (companies and the public sector), which were among the largest nonpayers of electricity bills.

- *Power theft and low collection rates.* These can be considered as subsidies because they lower the effective tariff rates paid by customers. It was estimated that 40 percent of electricity bills were uncollected in 1996, with the biggest nonpayers including other government-owned utility companies in water and heating.

- *Explicit budget support.* Although there were no direct subsidies for the Armenian power sector, state support was substantial in the form of loans made directly from the budget, which equaled, for example, 0.2 percent of GDP in 1996. In addition, the sector benefited from loan guarantees. Unpaid taxes were another source of support, equaling about 1.5 percent of GDP in 1996.

Reforms Since the Mid-1990s

Electricity prices rose sharply in the years 1995–99 toward cost-recovery levels, effectively eliminating the bulk of the subsidies. Residential tariffs more than doubled over 1995–99 to reach 25AMD per kWh, a level considered close to cost recovery (Figure 9.3) and consistent with the levels of tariffs charged to nonresidential users, considerably reducing cross-subsidies. Efficiency gains from the electricity sector reforms have helped reduce the fiscal burden of the sector in spite of a long period of unchanged prices since 1999. In 2009, retail prices were raised by 20 percent following an increase in the price of gas supplied by Russia.

Changes in both the level and structure of tariff increases were the centerpiece of the reforms. The tariff structure changed with the removal of the lifeline tariff in 1999. Although meant in principle to help protect low-income households by providing lower rates for low levels of consumption, the lifetime tariff was subject to abuse. In particular, households and meter readers colluded to delay the reporting of high winter consumption levels. A year earlier, discounted tariffs for low-

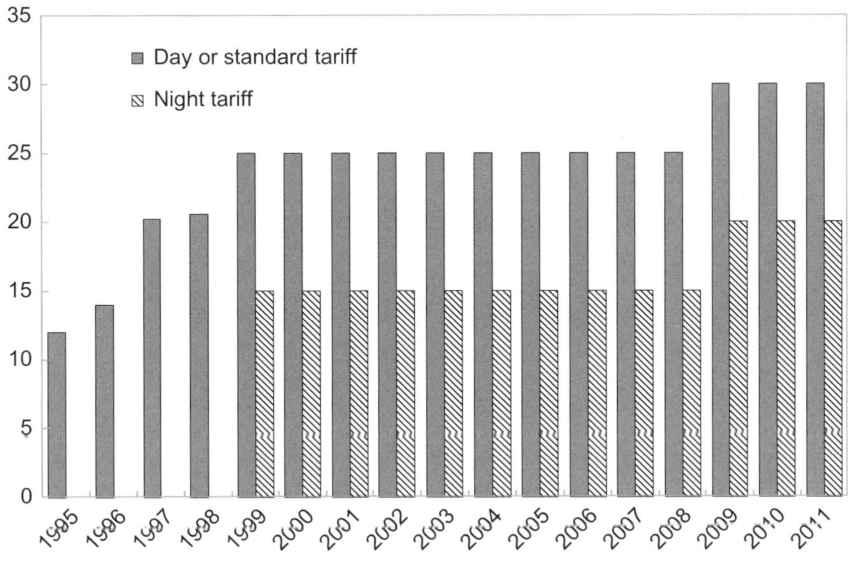

Source: National authorities.

Figure 9.3   Armenia: Residential Electricity Tariffs, 1995–2011 (Armenian drams per kWh)

income consumers, electricity company employees, and military personnel were withdrawn, following an overhaul of the social protection system to better target it to the poor. Price reforms contributed to reducing the deficit of the electricity sector from about 21 percent of GDP in 1994 to less than 3 percent of GDP in 2000 (Figure 9.4).

Efforts to improve collection rates also reduced the commercial losses of electricity companies. Meters were transferred from residential premises to communal hallways to prevent misreporting and tampering (Velody, Cain, and Philips, 2003). Meter readers were no longer allowed to collect cash, reducing the risks of corruption. Bill payments started to be collected through banks and post offices. Strict enforcement of disconnection policies has also led to an improvement in collection rates.

A public awareness campaign helped mobilize support for strengthening collections. The authorities emphasized that bill payments would help solve the problem of frequent power cuts and limited power availability (Velody, Cain, and Philips, 2003). The increase in collection rates was impressive, rising from 40 percent in 1996 to almost 100 percent by 2003, although it weakened temporarily in 1999 on account of the sharp increase in tariffs (Figure 9.5).

Tariff reforms were complemented by institutional reforms, paving the way for private sector participation. Private sector participation brought some efficiency gains, with system losses in percent of gross supply declining from 30 percent in 1999 to 10 percent in 2010. The authorities also established an independent regulator in 1997, with the mandate to set up and review electricity tariffs and regulate

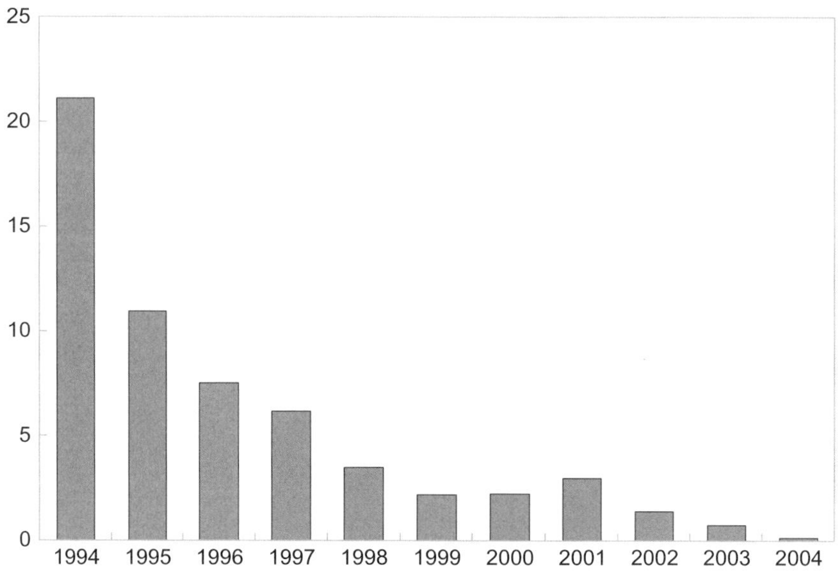

Source: Sargsyan, Balabanyan, and Hankinson (2006).

Figure 9.4 Armenia: Electricity Sector Financial Deficit, 1994–2004 (Percent of GDP)

the sector. The law empowers the regulator to ensure that tariffs fully cover medium-term costs, including depreciation, debt service, and other capital costs.

Strong political will and donor assistance supported subsidy reforms. According to Sargsyan, Balabanyan, and Hankinson (2006), political will was crucial for the initial impetus toward power sector reform and for the success of the privatization process. Despite the initial failed tender for privatizing the distribution system, the authorities learned from their early setbacks and persisted in their reform efforts, while also addressing weaknesses in the legal and regulatory framework. Further, the electricity price increase took place before privatization, demonstrating the authorities' commitment to reforms. Equally important was the fact that the government kept its commitments after privatization, notably by not backing down from the strict application of disconnection policies, even though some public organizations and ministries were affected. Finally, donors, including the IMF, the World Bank, and the United States Agency for International Development, provided significant support to the reform agenda, mainly through conditional loans and technical assistance.

Impact of the Reforms

The reform of the electricity sector contributed to fiscal adjustment. The fiscal deficit declined sharply—from 16.5 percent of GDP in 1994 to 9 percent of GDP in 1995 and further to 6.3 percent by 2000 (Figure 9.6).

The impact of electricity price increases on inflation was mitigated by successful macroeconomic stabilization. The pre-reform period (1993–94) was characterized

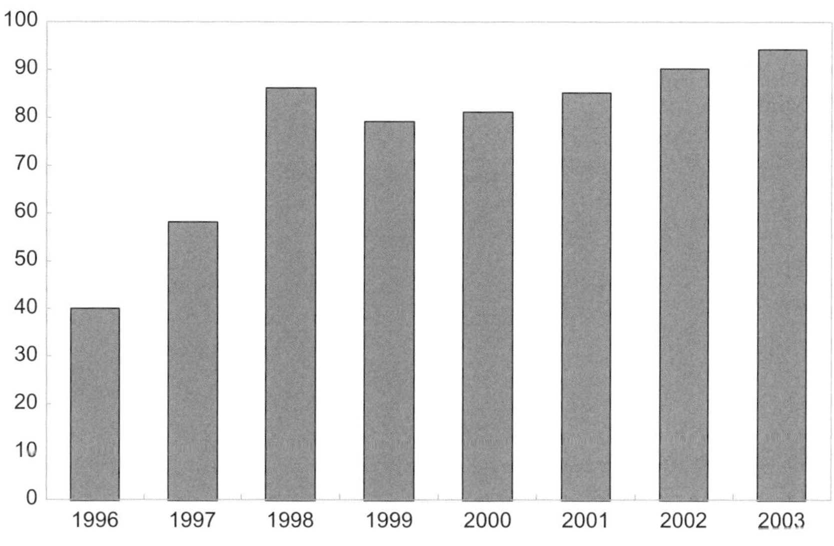

Source: Nixson and Walters (2005).

Figure 9.5 Armenia: Electricity Bill Collection Rate, 1996–2003 (Percent)

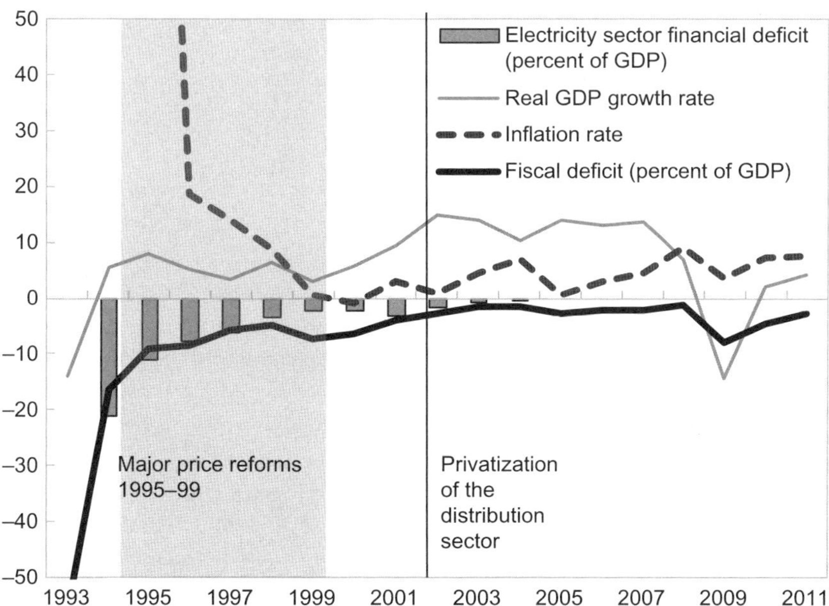

Sources: IEA; Sargsyan, Balabanyan, and Hankinson (2006); World Bank, *World Development Indicators*.

Figure 9.6 Armenia: Macroeconomic Developments and Electricity Subsidy Reforms, 1993–2011

by very high government deficits financed largely by the central bank. As a result, hyperinflation ensued. Successful macroeconomic stabilization resulted in a decline in inflation from over 5,000 percent in 1994 to single digits by 1998, reflecting monetary policy tightening and a sharp fiscal consolidation.

Sound macroeconomic policies and structural reforms were effective in facilitating growth, which averaged about 5.5 percent during 1995–99. It is difficult to disentangle the impact of the subsidy reform on growth from that of other macroeconomic policies and structural reforms, but it is likely that the electricity sector reform facilitated growth by improving power reliability and boosting electricity production.

Mitigating Measures

Offsetting measures were needed in light of the high share of the poor's expenditure on electricity. The 1999–2000 household survey showed that the share of electricity spending in household spending was almost twice as high in poor households as in nonpoor households (Table 9.3). This was especially the case among the urban poor. The following were the measures instituted:

- *Social safety net.* Tariff reforms coincided with an overhaul of the social safety nets, marked by the introduction of a cash transfer program, the Poverty Family Benefit. In 1999, the government replaced child and family allowances by a cash transfer program, the Poverty Family Benefit, which in contrast to the previous safety net program, is means tested. The program was not specifically targeted to offset the effect of higher electricity prices, but it has helped beneficiaries maintain real consumption in the face of higher electricity bills. The design of the benefit, however, helped increase the collection rate and improved energy efficiency, as the benefit is withdrawn if a household overconsumes and does not pay its electricity bill. The Poverty Family Benefit is considered a relatively well targeted program. It covered 25 percent of households initially, but coverage gradually declined to 18 percent in 2010 as eligibility criteria were tightened. This allowed an increase in the average payment by 40 percent in real terms between 2006, while maintaining the cost of the program at around 1 percent of GDP.

- *Cash transfers.* In addition, two one-off cash transfers were made to low-income households in 1999–2000 to help them cope with higher electricity prices. Beneficiaries included eligible households under the Poverty Family Benefit program and other households considered to have difficulties paying their bills.

- *Dual-rate meters.* A small-scale government program was started in 1999 to provide dual-rate electricity meters for low-income households. The use of dual-rate meters allowed households to benefit from discounted night tariffs and removed the need for energy suppliers to use high-cost generators during peak times of use.

TABLE 9.3

Armenia: Electricity Share in Total Household Spending *(Percent)*		
	Rural	**Urban**
Poor	13	16
Nonpoor	7	9

Source: Lampietti and others (2011).

Lessons

Strong political will is important for reform to succeed. The government was persistent in its efforts to achieve successful privatization and undertook politically costly tariff reforms. Although donors also played a role, their best efforts would have proven ineffective if government officials had not committed fully to the reforms (Sargsyan, Balabanyan, and Hankinson, 2006).

A good regulatory environment that limits interference in setting tariffs can facilitate reform. A proper legal framework for private sector participation was put in place, and an independent regulatory commission was created to determine tariffs.

Measures to improve collections are essential. Strict enforcement of disconnection policies and collection schemes (e.g., through bank accounts) that limit the risk of collusion between consumers and meter readers can increase collection rates.

An effective public awareness campaign linking payments of utility bills to more reliable service helps garner support for reform.

Implementation of mitigating measures for the poor helps fortify support for reform. A means-tested cash transfer program was introduced, improving the targeting of social safety nets. Additionally, two one-off cash payments and the provision of dual meters for low-income households helped soften the impact of electricity price increases on the poor, thereby facilitating public acceptance of the reforms. In the case of Armenia, there were useful synergies between the simultaneous reforms of the social protection system and the energy sector.

Turkey

Context

The Turkish electricity sector was dominated by a state-owned vertically integrated company prior to the reform. The Turkish Electricity Authority (TEK) was in control of generation, transmission, and distribution. TEK was later restructured into two separate state-owned companies, the Turkish Electricity Generation and Transmission Company (TEAS) and the Turkish Electricity Distribution Company (TEDAS).

The electricity sector reform started as part of the economy-wide reforms toward a market-oriented regime in the 1980s. The policy regime prior to these reforms was featured by heavy involvement of the state in economic activities, in particular in the form of government ownership of enterprises in critical industries, such as energy, telecommunications, petrochemicals, iron, and steel. The state also played a critical role in the allocation of financial resources, especially through state-owned banks. However, after a major balance of payments crisis in the second half of the 1970s and a military coup in 1980, Turkey was determined to transform its economy into a more market-oriented regime, through mass liberalization of domestic markets and international trade.

Reforms Since 1984

The electricity sector reform was set to achieve several objectives:

- *To better meet the growing electricity demand and to improve the fiscal position of the government.* The reform would eventually eliminate electricity subsidies, both consumer and producer subsidies. In addition, it was apparent that the government did not have the fiscal capacity to finance the expansions necessary to meet the future electricity demand.

- *To reduce the inefficiency in the electricity sector.* Private participation would introduce competition, improve efficiency, and limit monopoly abuse in the sector.

- *To meet the preconditions for Turkey's EU membership.* The reform had also been urged by various international institutions, which provided support during several economic crises.

Turkey has taken a series of steps to reform its electricity sector, with the goal to attract investment, encourage competition, and improve efficiency. The first law setting up a framework for private participation came into effect in 1984. The unbundling of the Turkish public electricity sector started in 1993. However, progress had been slow, with the public sector remaining dominant. An important attempt to privatize through sale of ownership rights in 1994 was struck down by the Constitutional Court. The privatization was able to resume only after an amendment to the Constitution in 1999. Instead, attempts to engage the private sector took the form of designing investment schemes such as build–operate–transfer, build–operate, and transfer-of-operating-rights contracts. These schemes, however, do not appear to have led to the development of competitive electricity markets in Turkey as these contracts locked generation companies into long-term exclusive sale agreements with predetermined prices and did not provide sufficient incentives for efficiency (Atiyas and Dutz, 2012).

By the end of the 1990s, the rapidly deteriorating fiscal stance led to pressures for a more ambitious privatization program, including in the electricity sector. In 2001, Turkey initiated a comprehensive electricity reform program by enacting its Electricity Market Law. The goal was to establish a competitive electricity market so as to increase private investment, improve efficiency, and ultimately strengthen Turkey's energy security while meeting the rapidly growing electricity demand. The state-owned enterprises were further unbundled into different business activities, including generation, transmission, distribution, and wholesale and retail

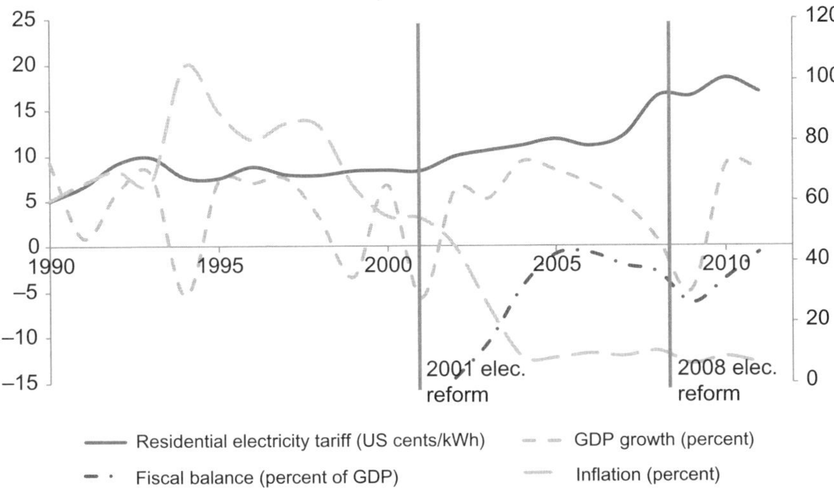

Sources: IEA; IMF staff estimates; IMF, WEO database.

Figure 9.7 Turkey: Macroeconomic Developments and Energy Subsidy Reforms, 1990–2010

supply. In 2006, a wholesale electricity market was also introduced to spur competition and improve efficiency.

Between 2002 and 2007, tariffs remained unchanged as demand rose. Despite the progress in restructuring the electricity sector, the tariffs for electricity remained unchanged between 2002 and 2007, although the prices of inputs had increased significantly. This disconnection between price and cost resulted in limited funding available for the maintenance of existing infrastructure and for new investment. In addition, the low electricity tariff contributed to the rapid rise in demand during this period.

To address these problems, the government started moving gradually to full cost recovery in the electricity sector in 2008. In January, electricity prices were increased by 20 percent from the fixed level in previous years. In March, the government approved a cost-based pricing mechanism that enables automatic quarterly tariff adjustments to cover the changes in the cost of supply. The new pricing mechanism became effective in July 2008 and resulted in several price increases by the end of 2009 (Figure 9.7). Although the electricity price increased more than 50 percent during this period, the impact on household welfare appears to have been limited because electricity consumption has accounted for only a relatively small share of the total household budget (Zhang, 2011).

Mitigating Measures

Turkey did not implement specific mitigating measures for the reform. It relied primarily on its social safety net to address the adverse impacts of electricity subsidy reforms on the poor.

Lessons

Broad support for and firm commitment to the market approach has been the key to progress of reform in the electricity sector in Turkey. The electricity sector reform started in the 1980s and has made significant progress despite several obstacles.

Improving economic conditions also helped advance reforms. A growing economy and improving standards of living assured the public that the country was moving in the right direction and helped move the reform forward.

Independent agencies for energy policy can help steer technical decisions away from politics. Under the Electricity Market Law, the Electricity Regulatory Market Authority (EMRA), the independent agency in charge of the petroleum sector, was also responsible for implementing the electricity market laws and regulating the electricity sector.

COAL SUBSIDIES

Poland

Context

In the pretransition era, coal mines were state owned and posed a substantial burden on the public finances. Coal mining was one of Poland's largest industries and employers, characterized by excess capacity and overemployment, which contributed—along with controlled prices—to operational deficits. The importance given to the coal mining industry, especially in the pretransition era, made the mining sector and its employees an economically and politically powerful lobby. This was also reflected in special privileges granted to coal sector employees, including free coal. Although there is no hard evidence available on the fiscal cost of maintaining the coal mining sector during the central planning period, data from the early transition period show that the sector ran operational deficits and had accumulated substantial debts.

Reforms Since 1990

In the 1990s, Poland started transforming its large and inefficient coal industry as part of the economic transition process. The government made several attempts to reform the sector with the aim to (1) close unprofitable mines; (2) reduce employment levels to improve labor productivity; (3) eliminate the sector's overcapacity; and (4) make the mining sector profitable, with the ultimate objective of privatizing mining companies. In a first restructuring program, which ran from 1990 through 1998, coal mines were transformed into state-owned enterprises (SOEs) and the SOEs were consolidated into seven coal companies.

However, these early attempts of reform showed only limited results in terms of reducing capacity, employment, and fiscal costs. This was mainly due to incomplete implementation of the reform agenda and resistance from unions to proposed wage cuts and reductions in employment. More specifically, the coal

TABLE 9.4

Poland: Key Macroeconomic Indicators, 1990–2011

	1990	1992	1994	1996	1993	2000	2003	2008	2010	2011
GDP per capita (US$)	1544.0	3149.1	6105.4	4056.0	4494.3	4477.7	5678.3	13876.3	12285.7	13539.8
GDP growth (percent)	−7.2	2.0	5.2	6.2	5.0	4.3	3.9	5.1	3.9	4.3
Inflation (percent)	585.8	43.0	32.2	19.9	11.8	10.1	0.8	4.2	2.5	4.3
Overall fiscal balance (GFS2001, percent of GDP)	n.a.	n.a.	n.a.	−4.9	−4.3	−3.0	−6.2	−3.7	−7.8	−5.2
Overall fiscal balance (GFS1986, percent of GDP)	0.0	−6.7	−2.9	−3.1	−2.5	−3.3	−5.6	−3.1	n.a.	n.a.
Public debt (GFS2001, percent of GDP)	n.a.	n.a.	n.a.	43.4	38.9	36.8	47.1	47.1	54.9	55.4
Public debt (GFS1986, percent of GDP)	90.1	82.4	64.6	42.4	36.7	37.7	48.4	47.0	n.a.	n.a.
Current account balance (percent of GDP)	1.9	1.0	5.3	−2.1	−4.0	−6.0	−2.5	−6.6	−4.7	−4.3
Oil imports (percent of GDP)	2.6	2.5	1.2	1.7	1.1	2.4	2.0	3.6	3.5	4.4
Oil exports (percent of GDP)	n.a.	n.a.	n.a.	n.a.	n.a.	n.a.	n.a.	n.a.	n.a.	n.a.
Oil consumption per capita (liters)	n.a.	n.a.	n.a.	n.a.	n.a.	407.6	401.5	530.0	470.7	n.a.
Coal production (million tons oil equivalent)	94.5	89.2	89.3	94.5	79.6	71.3	71.4	60.5	55.5	56.6
Coal consumption (million tons oil equivalent)	80.2	73.0	72.3	73.2	63.8	57.6	57.7	56.0	56.4	59.8
Coal price (Northwest Europe market in US$ per ton)	43.5	38.5	37.2	41.3	32.0	36.0	43.6	147.7	92.5	121.5
Poverty headcount ratio at US$1.25 per day (PPP) (percent of population)	n.a.	0.0	n.a.	1.4	0.1	0.1	n.a.	0.1	n.a.	n.a.

Sources: BP (2012); IMF, WEO database.
Note: GFS = Government Finance Statistics.

market was only gradually liberalized—allowing prices to increase to international levels by the mid-2000s (see Figure 9.8), limiting the opportunity for income growth for those mines that could have had viable operations under free-market conditions. In addition, the government provided insufficient resources to finance mine closures and social programs. As a result, the sector's debt level almost tripled between 1990 and 1998, amounting to US$5.6 billion (over 3 percent of GDP), despite significant transfers from the government and local authorities.

Only the new hard coal reform program, started in 1998, resulted in an effective restructuring of the Polish coal mining industry. The 1998–2002 hard coal reform—which fell into a less favorable economic situation than the previous reform attempt as GDP growth was on a downward trend and fiscal deficits increased (see Figure 9.9)—and several subsequent plans provided additional funding for social schemes and expressed a commitment to write off the debt that the mines had accumulated over past years. Under these plans, 21 uneconomic mines were closed, about 100,000 workers left the sector (Table 9.5), and about 70 percent of the coal mining industry's liabilities was written off—contributing to the 2003 spike in the fiscal deficit. The substantial reduction in employment and capacity allowed reducing production costs (see Figure 9.10), and the debt

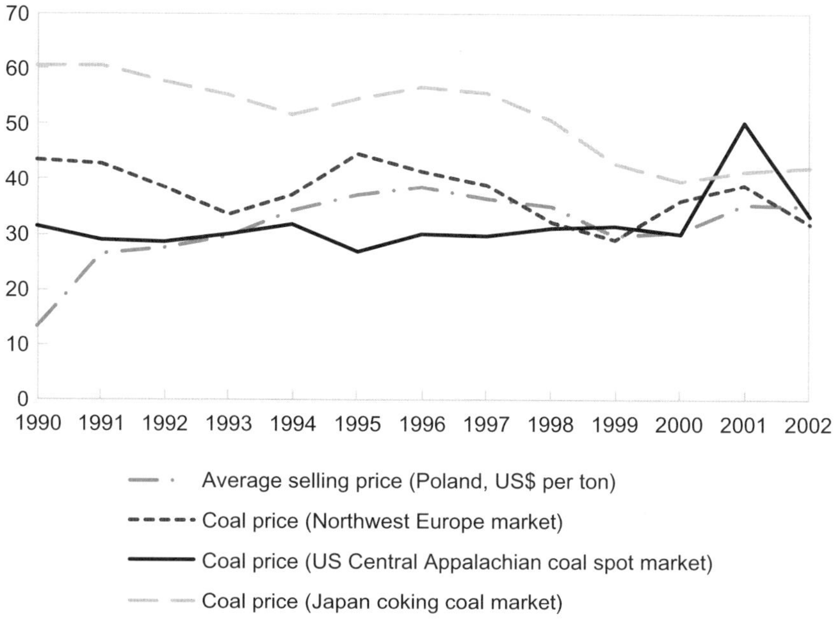

Sources: Blaschke and Lorenz (2004); BP (2012).

Figure 9.8 Poland: Coal Prices in Poland and Selected International Markets, 1990–2002 (U.S. dollars per ton)

reduction gave the industry the necessary financial freedom. Consequently, the sector has been profitable from 2003 onward, and a first privatization took place in 2009. With more decisive action from the government, closer cooperation with unions, and supporting programs from the European Development Bank and the World Bank, Poland's coal mining sector was transformed into a commercially viable industry. Today the Polish coal mining sector comprises 31 mines grouped into seven joint-stock holding companies and is dominated by three state-owned companies.

Mitigating Measures

The 1998 reform program was supported by social and labor market programs.

- *Social program.* The social program provided welfare benefits to dismissed workers while they transitioned into retirement or into new jobs. Under the social program of 1998–2002, more than 53,000 workers left coal mining, of which 33,000 received some form of help.

- *Labor market program.* The labor market program intended to redeploy especially younger coal workers elsewhere in the economy. It included soft loans for the establishment of businesses and services provided from newly established employment agencies, which offered training and other support to ease the transfer to other sectors.

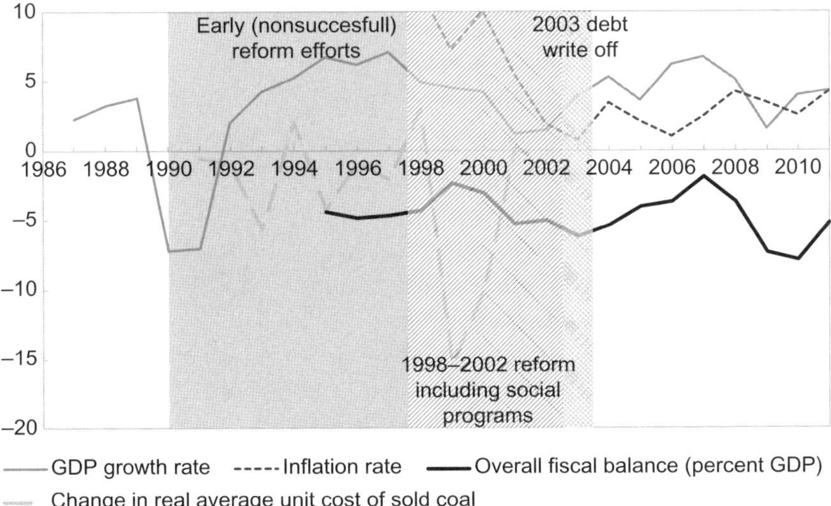

Sources: BP (2012); IMF, WEO.

Figure 9.9 Poland: Macroeconomic Developments and Coal Sector Reforms, 1987–2010 (Percent of GDP or Rate)

TABLE 9.5

Poland: Selected Indicators of Coal Mining Industry, 1990–2006

	1990	1992	1994	1996	1998	2000	2002	2004	2006
Number of operating mines	70	69	63	58	54	41	41	36	33
Production level (Mmt)	147	132	133	136	116	102	102	99	94
Employment (1,000 persons)	388	336	292	259	208	155	141	127	119
Productivity (mt per person)	380	392	454	526	558	659	725	780	790
Average coal price (US$ per mt)	37	45	52	45	41	38	38	53	57
Average coal production cost (US$ per mt)	54	53	51	48	49	37	37	44	55
Income (US$ million)	8,848	6,347	6,722	6,933	6,148	5,619	5,477	6,473	6,222
Production costs (US$ million)	8,104	7,237	7,233	7,731	7,760	6,107	5,634	5,568	6,025
Operating profit (US$ million)	744	−890	40	−798	−1,612	−488	−157	850	235
Net financial profit (US$ million)	−121	−1,497	−128	−777	−1,445	−504	−162	734	126
Debt (US$ million)	1,879	3,558	4,490	4,293	5,585	6,232	6,066	2,335	2,130
Total payments from government and local authorities (US$ million)	610	1,036	1,872	1,103	1,118	752	693	595	493

Source: Suwala (2010).

Note: Mmt = million metric tons; mt = metric ton.

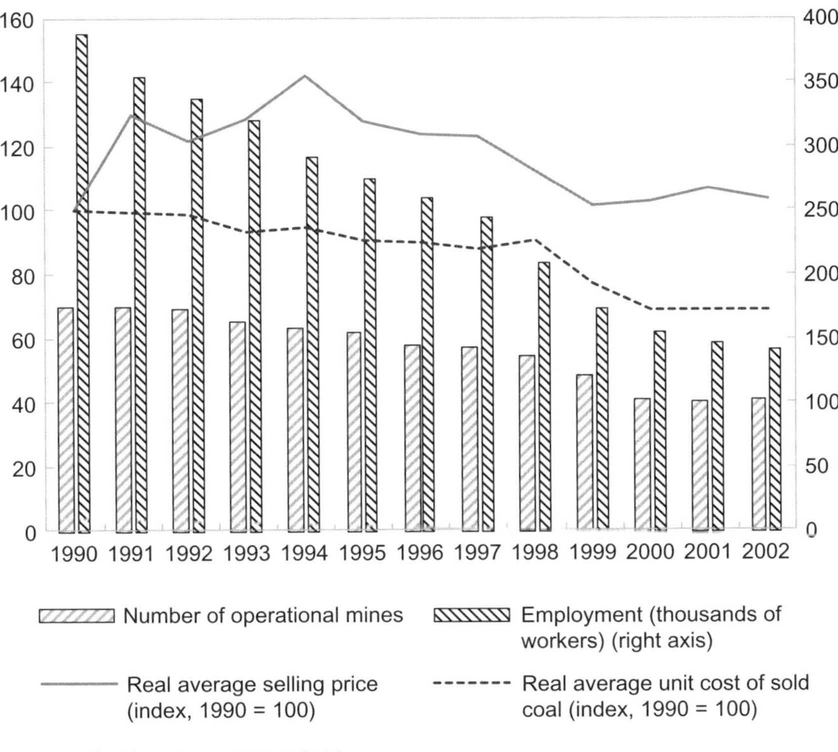

Sources: Blaschke and Lorenz (2004); BP (2012).

Figure 9.10   Poland: Indicators for the Coal Sector Reform, 1990–2002

Lessons

Reforms need political commitment and might also need some financial resources to complete. Not providing either or both of these might lead to a costly prolongation or even worsening of the drain of fiscal resources. When Poland made the first attempt to reform the coal mining sector, the government did not demonstrate full commitment to implementing the reforms, and it did not provide adequate funding for social programs. As a consequence, the reforms dragged on, and the sector continued running deficits and accumulating debt. The reform would have been less costly if it had been fully implemented from the beginning.

When reforming subsidies of nationally important industry, reforms need to cover all aspects of the industry, including product and labor markets. Coal mines in Poland could not become profitable until the coal market was liberalized and prices were able to adjust in line with international price fluctuations. In addition, preferential employment conditions in the coal mining sector made it difficult to motivate employees to leave the sector voluntarily. Converting the publically owned coal mines into stock companies that are managed according to business interests also leads to an adjustment of employment conditions. Together with the implementation of a social safety net for the sector's employees, this allowed for an increase in the mobility of the labor force and increased workers' willingness to move to other sectors.

Reforms that come with substantial employment losses in major industries have to be designed in cooperation with unions and need to be supported by appropriate social and labor market programs. In Poland, the first mining sector reform attempts were not successful because they did not provide adequate support for the miners, who were most affected by the reforms and who had a strong lobby. The mitigating measures designed in cooperation with the unions and included in subsequent reform plans broke the resistance of the miners to the restructuring. The case of Poland's coal industry demonstrated that the role of unions in the reform process and the magnitude of the required social and labor market programs are especially important for an industry that (1) is a major employer for the economy and the absolutely dominant employer in some regions and (2) has employees with very specialized skills that are of very limited use outside the industry.

The assumption of social liabilities and accumulated debt can be instrumental to successful subsidy reforms, especially when the sector needs to be modernized. To become profitable under market conditions, the Polish coal mining industry needed to be transformed from an oversized and inefficient energy provider for a centrally planned economy into an internationally competitive lean and modern industry. Given the substantial financial burden from the past—from the rehabilitation of old mines as well as from obligations toward the sector's employees—the sector could not have survived the reform without financial support from the government. The assumption of past liabilities, as well as substantial support for transition costs, allowed the industry to move toward profitability and to eventually be weaned from public support.

Appendix A. Estimating Pretax and Posttax Global Energy Subsidies

BAOPING SHANG, IAN PARRY, AND LOUIS SEARS

This appendix describes the data sources and methodologies used for the estimation of subsidies for petroleum products, coal, natural gas, and electricity.

PRETAX SUBSIDIES

Petroleum Products

Pretax consumer subsidies for gasoline, diesel, and kerosene are estimated as the difference between international prices, adjusted upward for transportation and distribution margins, and domestic consumer prices, for 176 countries between 2000 and 2011.[1] It should be noted that all these subsidy estimates are based on comparisons for prices of final products (e.g., gasoline) rather than unrefined products (such as crude oil).

For countries for which the Organisation for Economic Co-operation and Development (OECD) has detailed data on pretax prices and petroleum product taxes, pretax prices are used to measure supply costs. For other countries, supply costs are based on spot prices from the International Energy Agency (IEA). For net oil-importers among these countries, margins are assumed to be US$0.10 per liter to cover international transport costs and another US$0.10 per liter to cover domestic distribution and retailing costs. For net oil exporters, no adjustment is made, because the international transport cost is saved when the product is consumed domestically rather than exported. This is assumed to offset domestic distribution and retailing costs.

Domestic consumer prices for petroleum products (for both firms and households) are taken from publicly available sources for OECD countries. For other countries, domestic prices were provided by country authorities to IMF staff and supplemented by survey data from the Deutsche Gesellschaft für Internationale Zusammenarbeit (Ebert and others, 2009). For gasoline, the price is for regular unleaded or other grades, based on availability. Where consumer prices were

---

[1] Subsidies for oil-based heating fuels and fuels for non-road transportation vehicles, which are substantial in some countries, are not included because of data limitations.

unavailable, they were imputed based on observed pass-through behavior for that country. This was done for approximately 54 countries in 2009 and one country (Venezuela) in 2011. End-of-year prices are used to estimate subsidies except for 30 countries, mostly in the Middle East and North African (MENA) region, where quarterly price data are available. Producer subsidies for petroleum products are based on OECD producer support estimates (OECD, 2012a). These estimates capture both direct budgetary transfers and preferential treatment through the tax code to petroleum producers.

The fuel-product consumption levels used to calculate total subsidies are based on OECD and IEA data and include consumption by both households and enterprises.

Coal and Natural Gas

Consumer subsidy estimates are based on IEA data for coal in 39 countries and for natural gas in 37 countries between 2007 and 2011. IMF staff estimates on natural gas subsidies are available for an additional four countries in the MENA region. This calculation measures subsidies as the difference between the reference price and the domestic price paid by households and firms. The IEA reference prices for natural gas and coal, both traded goods, are defined differently for net importers and net exporters.[2] In addition, producer coal subsidies for 16 countries between 2007 and 2011 are based on OECD data.

For net importers, the reference price was defined as the price at the nearest international market, adjusted for quality differences, the cost of freight and insurance, distribution and marketing costs, and any value added tax (VAT). The price does not include excise duties. For net exporters, the reference price was calculated as the price at the nearest international market, adjusted for quality differences, less the costs of freight and insurance, plus distribution, marketing, and VAT. It should be noted that the quantities of coal and natural gas used in this calculation do not include the amount used for electricity and heat generation. To estimate pretax subsidies, the VAT is subtracted from the IEA estimates, using the standard VAT rate in the country. Producer subsidies for coal are based on OECD producer support estimates that capture the amount of tax subsidies (such as special income tax treatment) or budgetary expenditures designed to support producer incomes (OECD, 2012a).

Electricity

Given the varying availability of data, a number of different approaches are taken to measure subsidies. For 40 countries in sub-Saharan Africa, the Middle East and North Africa, and a few selected emerging economies in Europe, estimates of combined producer and consumer subsidies are compiled from various

---

[2]Although coal and natural gas are traded in international markets, transportation costs for these products are high, and a large share of both is consumed domestically or in the regions where they are produced.

World Bank reports and IMF staff estimates; thus, they are not necessarily comparable. For these countries, subsidy estimates are based on average domestic prices and cost-recovery prices that cover production costs, investment cost, distributional loss, and the nonpayment of electricity bills. An upward adjustment is also made for the input subsidies that electricity producers may receive through their use of subsidized fossil fuel products. For 31 of these 40 countries, the latest year for which data are available is 2009.

For 37 countries, estimates of consumer price subsidies between 2007 and 2011 are taken from the IEA, based on the difference between costs (adjusted for any subsidy on fossil fuel inputs) and average domestic prices (IEA, 2011b). As these prices do not include investment cost, nonpayment of electricity, or distributional losses, the estimates may understate the subsidies. In total, the sample covers 77 countries.

POSTTAX SUBSIDIES

Posttax subsidies are estimated as pretax subsidies plus

- *a corrective (or "Pigouvian") tax*, reflecting an (excise) tax on energy products to charge for externalities associated with CO_2 emissions, local pollution, and (in the case of gasoline and motor diesel) other externalities such as traffic congestion and accidents; and
- *a revenue component*, reflecting an (ad valorem) tax on energy products consumed by households that would be consistent with taxation of any other consumer good at the standard VAT or general sales tax (GST) rate.

Corrective Taxes

This section discusses the estimation of taxes needed to correct for externalities from petroleum products, coal, and natural gas. To avoid double counting we do not measure externalities from electricity generation, and because of the lack of available evidence, we do not measure externalities for other generation fuels.[3] Owing to the lack of systematic cross-country data, the subsidy estimates do not take into account certain charges and taxes for energy that are often rationalized on environmental grounds.[4] Environmental and transportation-related

---

[3] For example, for nuclear power it is extremely difficult to quantify the risks from radioactive waste and meltdowns.

[4] Principally, these include regional or country-level carbon pricing programs, road user charges (e.g., mileage-based tolls for trucks in Germany) and excise taxes on electricity consumption and vehicle sales. For example, the European Union Emissions Trading System imposes a carbon tax on certain greenhouse gases that are emitted by factories, power plants, and other installations in the system. The current prevailing prices, however, are only a small fraction of estimated damages, and only about half of emissions are covered under the scheme. To take another example, incorporating the road user charges for diesel vehicles in New Zealand would lower our posttax energy subsidy estimates for that country by US$0.8 billion.

externalities have been quantified for the United States and just a few other countries.[5]

Petroleum Products

Combustion of petroleum products (gasoline, diesel, and kerosene) generates both carbon dioxide (CO_2) emissions, which contribute to future global warming, and local air pollution, which elevates mortality risks for people inhaling the pollution. Other externalities associated with motor vehicle use—which we apportion to gasoline and diesel fuels—include traffic congestion and accidents and (primarily in the case of trucks) road damage. Table A.1 summarizes some estimates of motor fuel taxes to correct for these externalities that have been conducted for the United States, the United Kingdom, and Chile. The corrective tax estimate is highest for Chile, where there is a high incidence of traffic fatalities (especially for pedestrians), a large portion of nationwide driving occurs under highly congested conditions, and vehicles have relatively high emission rates.

For CO_2 emissions, we assume an illustrative value for global warming damages of US\$34 per ton (in 2007 dollars), following the United States Interagency Working Group on Social Cost of Carbon (2013). The estimates in the literature have varied considerably, however—for example, Nordhaus (2011) estimates damages of US\$12 per ton, and Stern (2006) estimates US\$85 per ton. The US\$34 per ton of CO_2 emissions is applied to all fuels and, for example, translates into US\$0.07–US\$0.09 per liter of gasoline or diesel (Table A.1).

A careful assessment of the noncarbon corrective fuel taxes for other countries would take into account a variety of local factors affecting the willingness to pay for reductions in these negative externalities, including, most important, income, local emission rates, population density, travel delays, and the frequency of traffic acci-

TABLE A.1

Corrective Motor Fuel Taxes, Selected Countries
(Cents per liter, 2011 dollars)

	Gasoline (cars)			Diesel (trucks)	
	United States	**United Kingdom**	**Chile**	**United States**	**Chile**
Total	38	44	73	40	65
Contribution of:					
local pollution	3	4	18	10	16
carbon	8	7	8	9	9
congestion	15	26	19	10	16
accidents	12	8	28	3	12
noise	0	0	0	2	1
road damage	0	0	0	6	12

Sources: Institute for Fiscal Studies (2012); Parry (2011); Parry and Small (2005); Parry and Strand (2011).
Note: The above studies estimate corrective diesel fuel taxes for the United States and Chile but not for the United Kingdom.

---

[5]More detailed work for other countries is under way in the Fiscal Affairs Department to provide more precise estimates (IMF, forthcoming).

dents. Internationally comprehensive data on these factors are not readily available, except for income per capita. We make adjustments to the estimates of willingness to pay by comparing a given country's income (e.g., Colombia) in purchasing parity terms with the United States, the United Kingdom, and Chile.[6] An income elasticity of 0.8 is assumed between the willingness to pay for reductions in externalities and per capita income, following the OECD (OECD, 2012b). We then apply this correction to the estimates of externalities per liter described in Appendix Table A.1 for the United States, the United Kingdom, and Chile. We then take the average across the three countries to arrive at our estimate for Colombia.

Coal

To estimate the corrective tax per ton of coal for global warming damages, we first derive CO_2 emissions per ton of coal, based on IEA data on coal consumption and CO_2 emissions from coal by country. The corrective tax per ton of coal is then calculated by multiplying CO_2 emissions per ton of coal consumption with the global warming damages of US\$34 per ton of CO_2 emissions.

Beyond its CO_2 emissions, the other major externality associated with coal combustion is local air pollution, where the most important problem is the fine particulates (that permeate the lungs) formed from chemical reactions involving sulfur dioxide (SO_2) emissions. A state-of-the-art modeling exercise for the United States by a committee of experts (National Research Council, 2009) put the local pollution damages from the average coal plant in 2005 at about US\$65 (in 2010 dollars) per (short) ton. This estimate is extrapolated to other countries based on per capita income, in the same way as for petroleum products. This approach does not adjust for cross-country differences in the pollution content of coal or the use of technologies to "scrub" emissions from the smokestack.

Natural Gas

Natural gas is far less emissions-intensive than coal—it produces about half the carbon emissions per unit of energy and only very minimal SO_2 emissions. For natural gas, then, only a carbon damage is applied. As with coal, the corrective tax is calculated (based on IEA data) by emissions per thousand cubic feet times US\$34 per ton of CO_2 emissions.

Revenue Component

Here a scenario where energy products would be taxed just like other consumer goods is considered.[7] The estimates are based on VAT rates for 150 countries in

---

[6]Posttax subsidies as a share of GDP for low-income countries would increase from 3.3 percent of GDP to 5.3 percent without this adjustment for noncarbon externalities of petroleum products and coal.

[7]In principle, individual products should be taxed more heavily, or less heavily, than the average consumer good (on revenue-raising grounds), depending on whether taxing them causes a significant shift toward untaxed goods (e.g., leisure and products that are exempt from VAT). However, there is little empirical support on which to make these types of adjustments, so they are not pursued here.

2011. For countries where VAT rates are not available or do not apply, the average VAT rate of countries with a similar level of income in the region is assumed.

Calculating Subsidies with Corrective Tax and Revenue Components

To quantify the magnitude of subsidies, the subsidy-free posttax prices are derived by applying the VAT or GST rates to both pretax prices and excise tax for externalities. The subsidy-free posttax prices are then compared with domestic prices and combined with consumption levels to compute subsidies. In the case of electricity, VAT or GST is estimated only for countries with pretax subsidies. This approach is followed because both domestic prices and cost-recovery prices are unknown for other countries. In the case of coal and natural gas, it is assumed that domestic prices in countries without pretax subsidies are the same as international reference prices.

One complication is that revenue from VAT would be effectively assessed only on energy products as final consumption goods, not on those as intermediate inputs for other consumption goods. To separate intermediate inputs from final consumption goods, we use IEA energy consumption data by industry type. It is assumed that energy products for residential use, commercial and public services, and gasoline for road use are final consumption goods. This approximation indicates that, on average, 99 percent of gasoline consumption, 7 percent of diesel consumption, 39 percent of kerosene consumption, 12 percent of coal consumption, 46 percent of natural gas consumption, and 51 percent of electricity consumption can be categorized as final consumption.

TABLE A.2

Pretax Subsidies in Percent of GDP for Petroleum Products, Electricity, Natural Gas, and Coal, 2011[1]
(Countries sorted by income category and region)

Country	Petroleum products	Electricity	Natural gas	Coal
Advanced				
Australia	0.04	n.a.	0.01	0.00
Austria	0.03	n.a.	n.a.	n.a.
Belgium	0.58	n.a.	n.a.	n.a.
Canada	0.05	n.a.	0.03	0.00
Cyprus	0.00	n.a.	n.a.	n.a.
Czech Republic	0.00	n.a.	n.a.	n.a.
Denmark	0.00	n.a.	n.a.	n.a.
Estonia	0.00	n.a.	n.a.	n.a.
Finland	0.00	n.a.	n.a.	n.a.
France	0.01	n.a.	0.00	n.a.
Germany	0.01	n.a.	0.00	0.07
Greece	0.09	n.a.	n.a.	0.00
Hong Kong SAR	0.00	n.a.	n.a.	n.a.
Iceland	0.00	n.a.	n.a.	n.a.
Ireland	0.00	n.a.	n.a.	0.05
Israel	0.00	n.a.	0.00	n.a.
Italy	0.00	n.a.	n.a.	n.a.
Japan	0.00	n.a.	0.00	n.a.
Korea	0.00	n.a.	0.00	0.02
Luxembourg	0.00	n.a.	n.a.	n.a.
Malta	0.00	n.a.	n.a.	n.a.
Netherlands	0.00	n.a.	n.a.	n.a.
New Zealand	0.00	n.a.	n.a.	n.a.
Norway	0.01	n.a.	0.01	0.00
Portugal	0.00	n.a.	n.a.	0.00
Singapore	0.00	n.a.	n.a.	n.a.
Slovak Republic	0.00	n.a.	n.a.	0.01
Slovenia	0.00	n.a.	n.a.	0.02
Spain	0.00	n.a.	n.a.	0.03
Sweden	0.00	n.a.	n.a.	n.a.
Switzerland	0.00	n.a.	n.a.	n.a.
Taiwan Province of China	n.a.	0.22	0.00	0.03
United Kingdom	0.01	n.a.	0.01	n.a.
United States	0.07	n.a.	0.02	0.00
CEE-CIS				
Albania	0.00	n.a.	n.a.	n.a.
Armenia	0.45	0.05	n.a.	n.a.
Azerbaijan	0.84	0.73	1.16	0.00
Belarus	0.00	0.26	n.a.	n.a.
Bosnia and Herzegovina	0.00	n.a.	n.a.	n.a.
Bulgaria	0.00	n.a.	n.a.	n.a.
Croatia	0.00	n.a.	n.a.	n.a.
Georgia	0.55	n.a.	n.a.	n.a.
Hungary	0.00	n.a.	n.a.	0.00
Kazakhstan	0.65	0.94	0.15	0.28
Kosovo	0.00	n.a.	n.a.	n.a.
Kyrgyz Republic	3.47	5.43	n.a.	n.a.
Latvia	0.00	n.a.	n.a.	n.a.
Lithuania	0.00	n.a.	n.a.	n.a.
Macedonia, FYR	0.00	n.a.	n.a.	n.a.
Moldova	0.00	n.a.	n.a.	n.a.

(Continued)

TABLE A.2 (CONTINUED)

Country	Petroleum products	Electricity	Natural gas	Coal
CEE-CIS (concluded)				
Mongolia	0.00	n.a.	n.a.	n.a.
Montenegro, Rep. of	0.00	n.a.	n.a.	n.a.
Poland	0.00	n.a.	n.a.	0.14
Romania	0.00	n.a.	n.a.	n.a.
Russia	0.00	0.99	1.09	0.00
Serbia	0.00	n.a.	n.a.	n.a.
Tajikistan	0.00	1.95	n.a.	n.a.
Turkey	0.00	n.a.	n.a.	0.02
Turkmenistan	6.00	2.32	14.80	n.a.
Ukraine	0.00	1.61	3.59	n.a.
Uzbekistan	0.02	5.71	18.88	n.a.
Emerging and Developing Asia				
Afghanistan	0.00	0.11	n.a.	n.a.
Bangladesh	0.90	2.63	1.60	0.00
Bhutan	0.51	n.a.	n.a.	n.a.
Brunei Darussalam	2.34	0.98	0.00	0.00
Cambodia	0.00	n.a.	n.a.	n.a.
China	0.00	0.15	n.a.	n.a.
Fiji	0.01	n.a.	n.a.	n.a.
India	1.25	0.32	0.17	0.00
Indonesia	2.58	0.66	0.00	0.00
Kiribati	n.a.	n.a.	n.a.	n.a.
Lao P.D.R.	0.00	n.a.	n.a.	n.a.
Malaysia	1.24	0.33	0.31	0.00
Maldives	0.19	n.a.	n.a.	n.a.
Myanmar	0.54	n.a.	n.a.	n.a.
Nepal	0.00	n.a.	n.a.	n.a.
Pakistan	0.13	1.31	2.54	0.00
Papua New Guinea	n.a.	n.a.	n.a.	n.a.
Philippines	0.00	0.00	0.00	0.00
Samoa	n.a.	n.a.	n.a.	n.a.
Solomon Islands	0.00	n.a.	n.a.	n.a.
Sri Lanka	1.16	0.47	0.00	0.00
Thailand	0.15	1.64	0.14	0.25
Timor-Leste	0.00	n.a.	n.a.	n.a.
Tonga	0.00	n.a.	n.a.	n.a.
Tuvalu	0.00	n.a.	n.a.	n.a.
Vanuatu	0.00	n.a.	n.a.	n.a.
Vietnam	0.00	2.38	0.13	n.a.
LAC				
Antigua and Barbuda	0.49	n.a.	n.a.	n.a.
Argentina	0.00	1.03	0.77	0.00
Bahamas, The	0.00	n.a.	n.a.	n.a.
Barbados	0.04	n.a.	n.a.	n.a.
Belize	0.00	n.a.	n.a.	n.a.
Bolivia	2.40	n.a.	n.a.	n.a.
Brazil	0.00	n.a.	n.a.	n.a.
Chile	0.00	0.00	0.00	0.00
Colombia	0.00	0.00	0.00	0.00
Costa Rica	0.00	n.a.	n.a.	n.a.
Dominica	0.00	n.a.	n.a.	n.a.
Dominican Republic	0.00	n.a.	n.a.	n.a.
Ecuador	6.31	0.18	0.00	0.00

TABLE A.2 (CONTINUED)

Country	Petroleum products	Electricity	Natural gas	Coal
LAC (concluded)				
El Salvador	0.00	0.00	0.00	0.00
Grenada	0.00	n.a.	n.a.	n.a.
Guatemala	0.00	n.a.	n.a.	n.a.
Guyana	0.00	n.a.	n.a.	n.a.
Haiti	n.a.	n.a.	n.a.	n.a.
Honduras	0.02	n.a.	n.a.	n.a.
Jamaica	0.00	n.a.	n.a.	n.a.
Mexico	0.00	0.00	0.00	0.00
Nicaragua	0.00	n.a.	n.a.	n.a.
Panama	0.02	n.a.	n.a.	n.a.
Paraguay	0.00	n.a.	n.a.	n.a.
Peru	0.00	0.00	0.00	0.00
St. Kitts and Nevis	0.20	n.a.	n.a.	n.a.
St. Lucia	0.19	n.a.	n.a.	n.a.
St. Vincent and the Grenadines	0.00	n.a.	n.a.	n.a.
Suriname	0.00	n.a.	n.a.	n.a.
Trinidad and Tobago	2.75	n.a.	n.a.	n.a.
Uruguay	0.00	n.a.	n.a.	n.a.
Venezuela	5.58	1.02	0.59	n.a.
MENA				
Algeria	4.30	1.08	5.36	0.00
Bahrain	5.37	2.57	n.a.	n.a.
Djibouti	0.00	0.45	n.a.	n.a.
Egypt	6.74	2.30	1.60	0.00
Iran	4.20	3.61	4.83	0.00
Iraq	9.92	1.39	0.25	0.00
Jordan	2.15	3.81	n.a.	n.a.
Kuwait	3.09	2.91	1.29	0.00
Lebanon	0.07	4.46	n.a.	n.a.
Libya	6.40	1.85	0.59	0.00
Mauritania	0.00	0.85	0.80	n.a.
Morocco	0.66	n.a.	n.a.	n.a.
Oman	3.01	0.76	2.20	n.a.
Qatar	1.22	1.20	1.07	0.00
Saudi Arabia	7.46	2.48	n.a.	0.00
Sudan	1.37	n.a.	n.a.	n.a.
Syria	n.a.	n.a.	n.a.	n.a.
Tunisia	0.77	2.23	n.a.	n.a.
United Arab Emirates	0.48	1.86	3.37	n.a.
Yemen	4.67	1.33	n.a.	n.a.
Sub-Saharan Africa				
Angola	1.30	0.27	0.00	0.00
Benin	0.00	1.78	n.a.	n.a.
Botswana	0.02	0.36	n.a.	n.a.
Burkina Faso	0.00	0.78	n.a.	n.a.
Burundi	0.00	n.a.	n.a.	n.a.
Cameroon	1.69	2.16	n.a.	n.a.
Cape Verde	0.00	2.17	n.a.	n.a.
Central African Republic	0.00	n.a.	n.a.	n.a.
Chad	0.00	0.00	n.a.	n.a.
Comoros	n.a.	n.a.	n.a.	n.a.
Congo, Democratic Republic of the	0.00	1.57	n.a.	n.a.

(Continued)

TABLE A.2 (CONCLUDED)

Country	Petroleum products	Electricity	Natural gas	Coal
Sub-Saharan Africa (concluded)				
Congo, Republic of	1.20	2.62	n.a.	n.a.
Côte d'Ivoire	0.00	2.72	n.a.	n.a.
Equatorial Guinea	0.28	n.a.	n.a.	n.a.
Eritrea	n.a.	n.a.	n.a.	n.a.
Ethiopia	0.19	1.24	n.a.	n.a.
Gabon	0.16	n.a.	n.a.	n.a.
Gambia, The	0.00	n.a.	n.a.	n.a.
Ghana	0.00	2.86	n.a.	n.a.
Guinea	0.00	n.a.	n.a.	n.a.
Guinea-Bissau	0.00	n.a.	n.a.	n.a.
Kenya	0.00	0.00	n.a.	n.a.
Lesotho	0.00	0.85	n.a.	n.a.
Liberia	0.00	n.a.	n.a.	n.a.
Madagascar	0.11	0.89	n.a.	n.a.
Malawi	0.00	1.60	n.a.	n.a.
Mali	0.00	0.93	n.a.	n.a.
Mauritius	0.00	n.a.	n.a.	n.a.
Mozambique	0.00	4.93	n.a.	n.a.
Namibia	0.00	0.52	n.a.	n.a.
Niger	0.00	0.00	n.a.	n.a.
Nigeria	1.42	1.31	0.00	0.00
Rwanda	0.00	0.29	n.a.	n.a.
Senegal	0.00	2.26	n.a.	n.a.
Seychelles	0.00	n.a.	n.a.	n.a.
Sierra Leone	0.00	n.a.	n.a.	n.a.
South Africa	0.01	0.55	0.00	0.00
Swaziland	0.00	n.a.	n.a.	n.a.
São Tomé and Príncipe	0.33	n.a.	n.a.	n.a.
Tanzania	0.00	2.10	n.a.	n.a.
Togo	0.00	n.a.	n.a.	n.a.
Uganda	0.00	1.32	n.a.	n.a.
Zambia	0.00	4.85	n.a.	n.a.
Zimbabwe	n.a.	14.52	n.a.	n.a.
World	**0.32**	**0.22**	**0.17**	**0.01**

Sources: Deutsche Gesellschaft für Internationale Zusammenarbeit (GIZ); International Energy Agency (IEA), *World Energy Outlook 2012;* IMF staff estimates; IMF, *World Economic Outlook;* Organisation for Economic Co-operation and Development (OECD); World Bank.

Note: Values are rounded to the nearest one-hundredth percent; electricity subsidies are taken from 2009 for 31 countries, and natural gas data are taken from 2010 for four countries. World estimates are calculated as identified subsidies divided by global GDP. n.a. = not applicable; CEE-CIS = Central and Eastern Europe and Commonwealth of Independent States; LAC = Latin America and the Caribbean; MENA = Middle East and North Africa.

[1]These subsidy estimates may differ from those in the country budget documents because of the methodologies described in this appendix.

TABLE A.3

Pretax Subsidies in Percent of Government Revenues for Petroleum Products, Electricity, Natural Gas, and Coal, 2011[1]
(Countries sorted by income category and region)

Country	Petroleum products	Electricity	Natural gas	Coal
Advanced				
Australia	0.13	n.a.	0.02	0.01
Austria	0.07	n.a.	n.a.	n.a.
Belgium	1.18	n.a.	n.a.	n.a.
Canada	0.13	n.a.	0.07	0.00
Cyprus	0.00	n.a.	n.a.	n.a.
Czech Republic	0.00	n.a.	n.a.	n.a.
Denmark	0.00	n.a.	n.a.	n.a.
Estonia	0.00	n.a.	n.a.	n.a.
Finland	0.00	n.a.	n.a.	n.a.
France	0.01	n.a.	0.00	n.a.
Germany	0.03	n.a.	0.00	0.17
Greece	0.23	n.a.	n.a.	0.01
Hong Kong SAR	0.00	n.a.	n.a.	n.a.
Iceland	0.00	n.a.	n.a.	n.a.
Ireland	0.00	n.a.	n.a.	0.14
Israel	0.00	n.a.	0.01	n.a.
Italy	0.00	n.a.	n.a.	n.a.
Japan	0.01	n.a.	0.00	n.a.
Korea	0.00	n.a.	0.00	0.06
Luxembourg	0.00	n.a.	n.a.	n.a.
Malta	0.00	n.a.	n.a.	n.a.
Netherlands	0.00	n.a.	n.a.	n.a.
New Zealand	0.00	n.a.	n.a.	n.a.
Norway	0.02	n.a.	0.02	0.00
Portugal	0.00	n.a.	n.a.	0.01
Singapore	0.00	n.a.	n.a.	n.a.
Slovak Republic	0.00	n.a.	n.a.	0.02
Slovenia	0.00	n.a.	n.a.	0.05
Spain	0.00	n.a.	n.a.	0.08
Sweden	0.00	n.a.	n.a.	n.a.
Switzerland	0.00	n.a.	n.a.	n.a.
Taiwan Province of China	n.a.	1.16	0.00	0.17
United Kingdom	0.03	n.a.	0.02	n.a.
United States	0.22	n.a.	0.06	0.01
CEE-CIS				
Albania	0.00	n.a.	n.a.	n.a.
Armenia	2.06	0.22	n.a.	n.a.
Azerbaijan	1.85	1.59	2.54	0.00
Belarus	0.00	0.62	n.a.	n.a.
Bosnia and Herzegovina	0.00	n.a.	n.a.	n.a.
Bulgaria	0.00	n.a.	n.a.	n.a.
Croatia	0.00	n.a.	n.a.	n.a.
Georgia	1.95	n.a.	n.a.	n.a.
Hungary	0.00	n.a.	n.a.	0.00
Kazakhstan	2.33	3.38	0.55	1.01
Kosovo	0.00	n.a.	n.a.	n.a.
Kyrgyz Republic	10.41	16.30	n.a.	n.a.
Latvia	0.00	n.a.	n.a.	n.a.
Lithuania	0.00	n.a.	n.a.	n.a.
Macedonia, FYR	0.00	n.a.	n.a.	n.a.
Moldova	0.00	n.a.	n.a.	n.a.

(Continued)

TABLE A.3 (CONTINUED)

Country	Petroleum products	Electricity	Natural gas	Coal
CEE-CIS (concluded)				
Mongolia	0.00	n.a.	n.a.	n.a.
Montenegro, Rep. of	0.00	n.a.	n.a.	n.a.
Poland	0.00	n.a.	n.a.	0.36
Romania	0.00	n.a.	n.a.	n.a.
Russia	0.00	2.58	2.85	0.00
Serbia	0.00	n.a.	n.a.	n.a.
Tajikistan	0.00	7.85	n.a.	n.a.
Turkey	0.00	n.a.	n.a.	0.07
Turkmenistan	31.84	12.29	78.48	n.a.
Ukraine	0.00	3.80	8.47	n.a.
Uzbekistan	0.06	14.20	46.94	n.a.
Emerging and Developing Asia				
Afghanistan	0.00	0.52	n.a.	n.a.
Bangladesh	7.56	22.12	13.45	0.00
Bhutan	1.39	n.a.	n.a.	n.a.
Brunei Darussalam	3.77	1.57	0.00	0.00
Cambodia	0.00	n.a.	n.a.	n.a.
China	0.00	0.68	n.a.	n.a.
Fiji	0.05	n.a.	n.a.	n.a.
India	6.75	1.72	0.90	0.00
Indonesia	14.51	3.69	0.00	0.00
Kiribati	n.a.	n.a.	n.a.	n.a.
Lao P.D.R.	0.00	n.a.	n.a.	n.a.
Malaysia	5.67	1.49	1.41	0.00
Maldives	0.61	n.a.	n.a.	n.a.
Myanmar	9.35	n.a.	n.a.	n.a.
Nepal	0.00	n.a.	n.a.	n.a.
Pakistan	1.02	10.23	19.89	0.00
Papua New Guinea	n.a.	n.a.	n.a.	n.a.
Philippines	0.00	0.00	0.00	0.00
Samoa	n.a.	n.a.	n.a.	n.a.
Solomon Islands	0.00	n.a.	n.a.	n.a.
Sri Lanka	7.99	3.26	0.00	0.00
Thailand	0.66	7.24	0.61	1.08
Timor-Leste	0.00	n.a.	n.a.	n.a.
Tonga	0.00	n.a.	n.a.	n.a.
Tuvalu	0.00	n.a.	n.a.	n.a.
Vanuatu	0.00	n.a.	n.a.	n.a.
Vietnam	0.00	8.59	0.47	n.a.
LAC				
Antigua and Barbuda	2.36	n.a.	n.a.	n.a.
Argentina	0.00	2.76	2.06	0.00
Bahamas, The	0.00	n.a.	n.a.	n.a.
Barbados	0.10	n.a.	n.a.	n.a.
Belize	0.00	n.a.	n.a.	n.a.
Bolivia	6.62	n.a.	n.a.	n.a.
Brazil	0.00	n.a.	n.a.	n.a.
Chile	0.00	0.00	0.00	0.00
Colombia	0.00	0.00	0.00	0.00
Costa Rica	0.00	n.a.	n.a.	n.a.
Dominica	0.00	n.a.	n.a.	n.a.
Dominican Republic	0.00	n.a.	n.a.	n.a.
Ecuador	15.44	0.44	0.00	0.00

TABLE A.3 (CONTINUED)

Country	Petroleum products	Electricity	Natural gas	Coal
LAC (concluded)				
El Salvador	0.00	0.00	0.00	0.00
Grenada	0.00	n.a.	n.a.	n.a.
Guatemala	0.00	n.a.	n.a.	n.a.
Guyana	0.00	n.a.	n.a.	n.a.
Haiti	n.a.	n.a.	n.a.	n.a.
Honduras	0.09	n.a.	n.a.	n.a.
Jamaica	0.00	n.a.	n.a.	n.a.
Mexico	0.00	0.00	0.00	0.00
Nicaragua	0.00	n.a.	n.a.	n.a.
Panama	0.08	n.a.	n.a.	n.a.
Paraguay	0.00	n.a.	n.a.	n.a.
Peru	0.00	0.00	0.00	0.00
St. Kitts and Nevis	0.55	n.a.	n.a.	n.a.
St. Lucia	0.68	n.a.	n.a.	n.a.
St. Vincent and the Grenadines	0.00	n.a.	n.a.	n.a.
Suriname	0.00	n.a.	n.a.	n.a.
Trinidad and Tobago	7.49	n.a.	n.a.	n.a.
Uruguay	0.00	n.a.	n.a.	n.a.
Venezuela	15.83	2.89	1.66	n.a.
MENA				
Algeria	10.84	2.72	13.52	0.00
Bahrain	18.96	9.08	n.a.	n.a.
Djibouti	0.00	1.32	n.a.	n.a.
Egypt	30.61	10.44	7.25	0.00
Iran	16.95	14.54	19.45	0.00
Iraq	12.69	1.78	0.32	0.00
Jordan	8.13	14.41	n.a.	n.a.
Kuwait	4.57	4.30	1.91	0.00
Lebanon	0.32	18.96	n.a.	n.a.
Libya	16.64	4.80	1.53	0.00
Mauritania	0.00	3.09	2.91	n.a.
Morocco	2.40	n.a.	n.a.	n.a.
Oman	7.28	1.83	5.31	n.a.
Qatar	3.17	3.12	2.78	0.00
Saudi Arabia	14.00	4.66	0.00	0.00
Sudan	7.33	n.a.	n.a.	n.a.
Syria	n.a.	n.a.	n.a.	n.a.
Tunisia	2.42	7.02	n.a.	n.a.
United Arab Emirates	1.38	5.32	9.61	n.a.
Yemen	19.03	5.42	n.a.	n.a.
Sub-Saharan Africa				
Angola	2.67	0.55	0.00	0.00
Benin	0.00	8.84	n.a.	n.a.
Botswana	0.07	1.21	n.a.	n.a.
Burkina Faso	0.00	3.59	n.a.	n.a.
Burundi	0.00	n.a.	n.a.	n.a.
Cameroon	8.92	11.42	n.a.	n.a.
Cape Verde	0.00	8.66	n.a.	n.a.
Central African Republic	0.00	n.a.	n.a.	n.a.
Chad	0.00	0.00	n.a.	n.a.
Comoros	n.a.	n.a.	n.a.	n.a.
Congo, Democratic Republic of the	0.00	5.75	n.a.	n.a.
Congo, Republic of	2.82	6.17	n.a.	n.a.

(Continued)

TABLE A.3 (CONCLUDED)

Country	Petroleum products	Electricity	Natural gas	Coal
Sub-Saharan Africa (concluded)				
Côte d'Ivoire	0.00	13.43	n.a.	n.a.
Equatorial Guinea	0.92	n.a.	n.a.	n.a.
Eritrea	n.a.	n.a.	n.a.	n.a.
Ethiopia	1.12	7.40	n.a.	n.a.
Gabon	0.56	n.a.	n.a.	n.a.
Gambia, The	0.00	n.a.	n.a.	n.a.
Ghana	0.00	14.70	n.a.	n.a.
Guinea	0.00	n.a.	n.a.	n.a.
Guinea-Bissau	0.00	n.a.	n.a.	n.a.
Kenya	0.00	0.00	n.a.	n.a.
Lesotho	0.00	1.61	n.a.	n.a.
Liberia	0.00	n.a.	n.a.	n.a.
Madagascar	0.95	7.86	n.a.	n.a.
Malawi	0.00	5.43	n.a.	n.a.
Mali	0.00	3.98	n.a.	n.a.
Mauritius	0.00	n.a.	n.a.	n.a.
Mozambique	0.00	16.40	n.a.	n.a.
Namibia	0.00	1.82	n.a.	n.a.
Niger	0.00	0.00	n.a.	n.a.
Nigeria	4.82	4.44	0.00	0.00
Rwanda	0.00	1.14	n.a.	n.a.
Senegal	0.00	10.08	n.a.	n.a.
Seychelles	0.00	n.a.	n.a.	n.a.
Sierra Leone	0.00	n.a.	n.a.	n.a.
South Africa	0.02	2.01	0.00	0.00
Swaziland	0.00	n.a.	n.a.	n.a.
São Tomé and Príncipe	0.90	n.a.	n.a.	n.a.
Tanzania	0.00	9.50	n.a.	n.a.
Togo	0.00	n.a.	n.a.	n.a.
Uganda	0.00	8.95	n.a.	n.a.
Zambia	0.00	21.59	n.a.	n.a.
Zimbabwe	n.a.	47.02	n.a.	n.a.
World	**0.94**	**0.64**	**0.50**	**0.03**

Sources: GIZ; IEA, *World Energy Outlook 2012;* IMF staff estimates; IMF, *World Economic Outlook;* OECD; World Bank.
Note: Values are rounded to the nearest one-hundredth percent; electricity subsidies are taken from 2009 for 31 countries, and natural gas data are taken from 2010 for four countries. World estimates are calculated as identified subsidies divided by global government revenues. n.a. = not applicable; CEE-CIS = Central and Eastern Europe and Commonwealth of Independent States; LAC = Latin America and the Caribbean; MENA = Middle East and North Africa.
[1]These subsidy estimates may differ from those in the country budget documents because of the methodologies described in this appendix.

TABLE A.4

Posttax Subsidies as Percent of GDP for Petroleum Products, Electricity, Natural Gas, and Coal, 2011[1]
(Countries sorted by income category and region)

Country	Petroleum products	Electricity	Natural gas	Coal
Advanced				
Australia	0.61	n.a.	0.19	0.68
Austria	0.04	n.a.	0.17	0.20
Belgium	0.58	n.a.	0.29	0.11
Canada	1.16	n.a.	0.47	0.26
Cyprus	0.09	n.a.	n.a.	0.01
Czech Republic	0.00	n.a.	0.37	1.75
Denmark	0.00	n.a.	0.11	0.22
Estonia	0.00	n.a.	0.21	3.34
Finland	0.00	n.a.	0.11	0.42
France	0.01	n.a.	0.14	0.08
Germany	0.01	n.a.	0.19	0.56
Greece	0.09	n.a.	0.11	0.58
Hong Kong SAR	0.42	n.a.	0.12	0.85
Iceland	0.00	n.a.	n.a.	0.14
Ireland	0.00	n.a.	0.19	0.27
Israel	0.00	n.a.	0.15	0.69
Italy	0.00	n.a.	0.31	0.14
Japan	0.10	n.a.	0.17	0.41
Korea	0.06	n.a.	0.34	1.55
Luxembourg[2]	2.62	n.a.	0.17	0.03
Malta	0.00	n.a.	n.a.	n.a.
Netherlands	0.00	n.a.	0.42	0.20
New Zealand	0.77	n.a.	0.18	0.20
Norway	0.01	n.a.	0.11	0.04
Portugal	0.00	n.a.	0.17	0.19
Singapore	0.49	n.a.	0.27	0.01
Slovak Republic	0.00	n.a.	0.50	0.79
Slovenia	0.00	n.a.	0.13	0.64
Spain	0.00	n.a.	0.18	0.21
Sweden	0.00	n.a.	0.02	0.09
Switzerland	0.00	n.a.	0.04	0.01
Taiwan Province of China	n.a.	0.28	0.25	2.06
United Kingdom	0.01	n.a.	0.32	0.28
United States	1.58	n.a.	0.36	0.78
CEE-CIS				
Albania	0.00	n.a.	0.01	0.02
Armenia	0.93	0.40	1.19	n.a.
Azerbaijan	2.39	0.91	2.19	0.00
Belarus	0.00	1.08	3.54	n.a.
Bosnia and Herzegovina	0.00	n.a.	0.11	4.66
Bulgaria	0.00	n.a.	0.38	2.91
Croatia	0.00	n.a.	0.45	0.29
Georgia	0.86	n.a.	0.61	0.08
Hungary	0.00	n.a.	0.78	0.39
Kazakhstan	2.36	0.97	1.32	3.64
Kosovo	0.00	n.a.	n.a.	0.02
Kyrgyz Republic	7.28	5.71	0.40	1.94
Latvia	0.00	n.a.	0.58	0.14
Lithuania	0.00	n.a.	0.55	0.14
Macedonia, FYR	0.00	n.a.	0.13	1.77
Moldova	0.00	n.a.	2.16	0.17

(Continued)

TABLE A.4 (CONTINUED)

Country	Petroleum products	Electricity	Natural gas	Coal
CEE-CIS (concluded)				
Mongolia	0.00	n.a.	n.a.	6.36
Montenegro, Rep. of	0.00	n.a.	n.a.	0.00
Poland	0.00	n.a.	0.26	2.33
Romania	0.00	n.a.	0.58	0.74
Russia	1.64	1.27	3.07	1.36
Serbia	0.00	n.a.	0.52	3.33
Tajikistan	0.16	2.50	0.32	0.20
Turkey	0.00	n.a.	0.45	0.86
Turkmenistan	8.59	2.39	21.94	n.a.
Ukraine	0.31	1.85	8.24	3.71
Uzbekistan	1.10	5.95	28.03	0.38
Emerging and Developing Asia				
Afghanistan	0.14	0.19	n.a.	n.a.
Bangladesh	1.46	3.01	2.94	0.12
Bhutan	1.36	n.a.	n.a.	n.a.
Brunei Darussalam	6.06	1.36	1.64	0.00
Cambodia	0.00	n.a.	n.a.	0.00
China	0.00	0.30	0.13	4.41
Fiji	0.16	n.a.	n.a.	n.a.
India	2.02	0.36	0.41	2.63
Indonesia	3.47	0.72	0.44	0.65
Kiribati	n.a.	n.a.	n.a.	n.a.
Lao P.D.R.	0.00	n.a.	n.a.	n.a.
Malaysia	5.38	0.56	1.02	0.98
Maldives	1.74	n.a.	n.a.	n.a.
Myanmar	1.04	n.a.	n.a.	n.a.
Nepal	0.28	n.a.	n.a.	0.16
Pakistan	1.14	1.63	3.67	0.22
Papua New Guinea	n.a.	n.a.	n.a.	n.a.
Philippines	0.31	0.00	0.11	0.65
Samoa	n.a.	n.a.	n.a.	n.a.
Solomon Islands	0.00	n.a.	n.a.	n.a.
Sri Lanka	2.17	0.75	0.00	0.04
Thailand	1.54	1.76	0.99	1.06
Timor-Leste	0.05	n.a.	n.a.	n.a.
Tonga	0.00	n.a.	n.a.	n.a.
Tuvalu	0.00	n.a.	n.a.	n.a.
Vanuatu	0.00	n.a.	n.a.	n.a.
Vietnam	0.83	2.64	0.78	1.60
LAC				
Antigua and Barbuda	1.77	n.a.	n.a.	n.a.
Argentina	0.35	1.15	1.56	0.13
Bahamas, The	1.57	n.a.	n.a.	n.a.
Barbados	0.61	n.a.	n.a.	n.a.
Belize	0.00	n.a.	n.a.	n.a.
Bolivia	5.18	n.a.	1.02	n.a.
Brazil	0.11	n.a.	0.11	0.10
Chile	1.36	0.00	0.12	0.42
Colombia	0.00	0.00	0.25	0.27
Costa Rica	0.47	n.a.	n.a.	0.03
Dominica	1.30	n.a.	n.a.	n.a.
Dominican Republic	0.06	n.a.	0.17	0.18
Ecuador	10.03	0.33	0.07	0.00
El Salvador	0.90	0.00	0.00	0.00

TABLE A.4 (CONTINUED)

Country	Petroleum products	Electricity	Natural gas	Coal
LAC (concluded)				
Grenada	1.12	n.a.	n.a.	n.a.
Guatemala	0.87	n.a.	n.a.	0.48
Guyana	1.20	n.a.	n.a.	n.a.
Haiti	n.a.	n.a.	n.a.	n.a.
Honduras	0.67	n.a.	n.a.	0.01
Jamaica	0.60	n.a.	n.a.	0.06
Mexico	2.14	0.00	0.42	0.16
Nicaragua	0.11	n.a.	n.a.	n.a.
Panama	2.40	n.a.	n.a.	0.02
Paraguay	0.00	n.a.	n.a.	n.a.
Peru	0.33	0.00	0.36	0.04
St. Kitts and Nevis	1.35	n.a.	n.a.	n.a.
St. Lucia	0.99	n.a.	n.a.	n.a.
St. Vincent and the Grenadines	0.99	n.a.	n.a.	n.a.
Suriname	0.00	n.a.	n.a.	n.a.
Trinidad and Tobago	5.95	n.a.	6.51	n.a.
Uruguay	0.00	n.a.	0.02	0.00
Venezuela	8.31	1.27	1.25	0.00
MENA				
Algeria	6.31	1.15	6.37	0.00
Bahrain	10.26	2.96	2.74	n.a.
Djibouti	0.24	0.51	n.a.	n.a.
Egypt	8.84	2.50	3.05	0.07
Iran	7.98	3.64	7.10	0.02
Iraq	14.91	1.57	0.34	0.00
Jordan	5.65	4.10	0.50	n.a.
Kuwait	7.01	3.12	2.05	0.00
Lebanon	3.93	4.61	0.24	0.15
Libya	9.11	2.33	1.90	0.00
Mauritania	1.06	0.93	0.80	n.a.
Morocco	3.04	n.a.	0.05	0.46
Oman	6.73	0.94	3.86	n.a.
Qatar	5.51	1.26	2.08	0.00
Saudi Arabia	13.57	2.79	0.96	0.00
Sudan	2.40	n.a.	n.a.	n.a.
Syria	n.a.	n.a.	n.a.	n.a.
Tunisia	2.75	2.43	1.01	n.a.
United Arab Emirates	3.58	2.04	4.67	0.04
Yemen	7.25	1.47	1.53	n.a.
Sub-Saharan Africa				
Angola	2.65	0.31	0.05	0.00
Benin	0.47	2.01	n.a.	n.a.
Botswana	1.05	0.48	n.a.	0.45
Burkina Faso	0.33	0.94	n.a.	n.a.
Burundi	0.00	n.a.	n.a.	n.a.
Cameroon	2.57	2.41	0.07	n.a.
Cape Verde	0.00	2.57	n.a.	n.a.
Central African Republic	0.00	n.a.	n.a.	n.a.
Chad	0.00	0.02	n.a.	n.a.
Comoros	n.a.	n.a.	n.a.	n.a.
Congo, Democratic Republic of the	0.00	1.80	0.00	0.13
Congo, Republic of	2.21	2.66	0.01	n.a.
Côte d'Ivoire	0.00	2.96	0.56	n.a.
Equatorial Guinea	2.02	n.a.	n.a.	n.a.

(Continued)

TABLE A.4 (CONCLUDED)

Country	Petroleum products	Electricity	Natural gas	Coal
Sub-Saharan Africa (concluded)				
Eritrea	n.a.	n.a.	n.a.	n.a.
Ethiopia	0.68	1.32	n.a.	n.a.
Gabon	0.88	n.a.	0.09	n.a.
Gambia, The	0.00	n.a.	n.a.	n.a.
Ghana	0.59	3.02	n.a.	n.a.
Guinea	0.00	n.a.	n.a.	n.a.
Guinea-Bissau	0.00	n.a.	n.a.	n.a.
Kenya	0.61	0.16	n.a.	0.01
Lesotho	0.05	0.94	n.a.	n.a.
Liberia	0.02	n.a.	n.a.	n.a.
Madagascar	0.45	0.98	n.a.	n.a.
Malawi	0.17	2.01	n.a.	n.a.
Mali	0.19	0.99	n.a.	n.a.
Mauritius	0.00	n.a.	n.a.	n.a.
Mozambique	0.27	5.07	0.13	0.02
Namibia	0.13	0.52	n.a.	0.10
Niger	0.24	0.17	n.a.	n.a.
Nigeria	2.16	1.34	0.28	0.00
Rwanda	0.00	0.39	n.a.	n.a.
Senegal	0.00	2.51	0.01	0.23
Seychelles	0.00	n.a.	n.a.	n.a.
Sierra Leone	0.63	n.a.	n.a.	n.a.
South Africa	0.40	0.73	0.00	3.24
Swaziland	0.00	n.a.	n.a.	n.a.
São Tomé and Príncipe	0.63	n.a.	n.a.	n.a.
Tanzania	0.00	2.26	0.27	0.04
Togo	0.85	n.a.	n.a.	n.a.
Uganda	0.00	1.45	n.a.	n.a.
Zambia	0.00	4.96	n.a.	0.00
Zimbabwe	n.a.	14.89	n.a.	3.08
World	**1.04**	**0.26**	**0.54**	**1.02**

Sources: GIZ; IEA, *World Energy Outlook 2012;* IMF staff estimates; IMF, *World Economic Outlook;* OECD; World Bank.
Note: Values are rounded to the nearest one-hundredth percent; electricity subsidies are taken from 2009 for 31 countries, and natural gas data are taken from 2010 for four countries. World estimates are calculated as identified subsidies divided by global GDP. n.a. = not applicable; CEE-CIS = Central and Eastern Europe and Commonwealth of Independent States; LAC = Latin America and the Caribbean; MENA = Middle East and North Africa.
[1]Owing to the lack of systematic cross-country data, the subsidy estimates do not take into account certain charges and taxes for energy that are often rationalized on environmental grounds. Principally, these include regional or country-level carbon pricing programs, road user charges (e.g., mileage-based tolls for trucks in Germany), and excise taxes on electricity consumption and vehicle sales. For example, the European Union Emissions Trading System imposes a carbon tax on certain greenhouse gases that are emitted by factories, power plants, and other installations in the system. The current prevailing prices, however, are only a small fraction of estimated damages, and only about half of emissions are covered under the scheme. To take another example, incorporating the road user charges for diesel vehicles in New Zealand would lower our posttax energy subsidy estimates for that country by US$0.8 billion.
[2]The estimate for Luxembourg reflects, to a large extent, cross-border sales of petroleum products to neighboring countries, with buyers attracted by lower tax rates.

TABLE A.5

Posttax Subsidies in Percent of Government Revenues for Petroleum Products, Electricity, Natural Gas, and Coal, 2011[1]
(Countries sorted by income category and region)

Country	Petroleum products	Electricity	Natural gas	Coal
Advanced				
Australia	1.90	n.a.	0.60	2.14
Austria	0.07	n.a.	0.36	0.42
Belgium	1.18	n.a.	0.58	0.23
Canada	3.04	n.a.	1.22	0.68
Cyprus	0.22	n.a.	n.a.	0.02
Czech Republic	0.00	n.a.	0.91	4.33
Denmark	0.00	n.a.	0.21	0.40
Estonia	0.00	n.a.	0.48	7.57
Finland	0.00	n.a.	0.20	0.79
France	0.01	n.a.	0.27	0.16
Germany	0.03	n.a.	0.42	1.25
Greece	0.23	n.a.	0.28	1.41
Hong Kong SAR	1.74	n.a.	0.48	3.50
Iceland	0.00	n.a.	n.a.	0.33
Ireland	0.00	n.a.	0.55	0.79
Israel	0.00	n.a.	0.38	1.70
Italy	0.00	n.a.	0.68	0.31
Japan	0.34	n.a.	0.54	1.33
Korea	0.25	n.a.	1.46	6.63
Luxembourg[2]	6.33	n.a.	0.41	0.08
Malta	0.00	n.a.	n.a.	n.a.
Netherlands	0.00	n.a.	0.93	0.45
New Zealand	2.66	n.a.	0.62	0.68
Norway	0.02	n.a.	0.19	0.08
Portugal	0.00	n.a.	0.38	0.43
Singapore	1.97	n.a.	1.10	0.05
Slovak Republic	0.00	n.a.	1.54	2.43
Slovenia	0.00	n.a.	0.32	1.53
Spain	0.00	n.a.	0.51	0.58
Sweden	0.00	n.a.	0.04	0.19
Switzerland	0.00	n.a.	0.12	0.02
Taiwan Province of China	n.a.	1.48	1.34	10.94
United Kingdom	0.03	n.a.	0.86	0.76
United States	5.03	n.a.	1.16	2.49
CEE-CIS				
Albania	0.00	n.a.	0.06	0.07
Armenia	4.25	1.81	5.44	n.a.
Azerbaijan	5.25	2.00	4.81	0.00
Belarus	0.00	2.58	8.43	n.a.
Bosnia and Herzegovina	0.00	n.a.	0.23	10.03
Bulgaria	0.00	n.a.	1.17	8.98
Croatia	0.00	n.a.	1.23	0.79
Georgia	3.04	n.a.	2.16	0.27
Hungary	0.00	n.a.	1.48	0.73
Kazakhstan	8.51	3.49	4.74	13.09
Kosovo	0.00	n.a.	n.a.	0.06
Kyrgyz Republic	21.84	17.13	1.19	5.83
Latvia	0.00	n.a.	1.62	0.38
Lithuania	0.00	n.a.	1.68	0.42
Macedonia, FYR	0.00	n.a.	0.44	6.16
Moldova	0.00	n.a.	5.89	0.46

(Continued)

TABLE A.5 (CONTINUED)

Country	Petroleum products	Electricity	Natural gas	Coal
CEE-CIS (concluded)				
Mongolia	0.00	n.a.	n.a.	16.02
Montenegro, Rep. of	0.00	n.a.	n.a.	0.00
Poland	0.00	n.a.	0.68	6.06
Romania	0.00	n.a.	1.84	2.34
Russia	4.27	3.30	8.01	3.54
Serbia	0.00	n.a.	1.27	8.12
Tajikistan	0.65	10.04	1.28	0.81
Turkey	0.00	n.a.	1.29	2.48
Turkmenistan	45.59	12.67	116.38	n.a.
Ukraine	0.73	4.36	19.45	8.76
Uzbekistan	2.72	14.80	69.68	0.94
Emerging and Developing Asia				
Afghanistan	0.66	0.86	n.a.	n.a.
Bangladesh	12.27	25.25	24.72	1.02
Bhutan	3.72	n.a.	n.a.	n.a.
Brunei Darussalam	9.73	2.19	2.64	0.00
Cambodia	0.00	n.a.	n.a.	0.01
China	0.00	1.34	0.58	19.45
Fiji	0.62	n.a.	n.a.	n.a.
India	10.91	1.97	2.21	14.17
Indonesia	19.46	4.04	2.45	3.65
Kiribati	n.a.	n.a.	n.a.	n.a.
Lao P.D.R.	0.00	n.a.	n.a.	n.a.
Malaysia	24.61	2.54	4.66	4.46
Maldives	5.57	n.a.	n.a.	n.a.
Myanmar	18.11	n.a.	n.a.	n.a.
Nepal	1.57	n.a.	n.a.	0.89
Pakistan	8.93	12.76	28.68	1.72
Papua New Guinea	n.a.	n.a.	n.a.	n.a.
Philippines	1.80	0.00	0.64	3.74
Samoa	n.a.	n.a.	n.a.	n.a.
Solomon Islands	0.00	n.a.	n.a.	n.a.
Sri Lanka	14.96	5.17	0.00	0.27
Thailand	6.80	7.77	4.38	4.68
Timor-Leste	0.06	n.a.	n.a.	n.a.
Tonga	0.00	n.a.	n.a.	n.a.
Tuvalu	0.00	n.a.	n.a.	n.a.
Vanuatu	0.00	n.a.	n.a.	n.a.
Vietnam	3.00	9.54	2.81	5.78
LAC				
Antigua and Barbuda	8.58	n.a.	n.a.	n.a.
Argentina	0.95	3.08	4.18	0.34
Bahamas, The	8.75	n.a.	n.a.	n.a.
Barbados	1.70	n.a.	n.a.	n.a.
Belize	0.00	n.a.	n.a.	n.a.
Bolivia	14.31	n.a.	2.83	n.a.
Brazil	0.33	n.a.	0.30	0.28
Chile	5.49	0.00	0.50	1.70
Colombia	0.00	0.00	0.93	1.00
Costa Rica	3.38	n.a.	n.a.	0.19
Dominica	4.18	n.a.	n.a.	n.a.
Dominican Republic	0.45	n.a.	1.24	1.34

TABLE A.5 (CONTINUED)

Country	Petroleum products	Electricity	Natural gas	Coal
LAC (concluded)				
Ecuador	24.55	0.80	0.18	0.00
El Salvador	5.05	0.00	0.00	0.00
Grenada	5.02	n.a.	n.a.	n.a.
Guatemala	7.33	n.a.	n.a.	4.09
Guyana	4.33	n.a.	n.a.	n.a.
Haiti	n.a.	n.a.	n.a.	n.a.
Honduras	2.85	n.a.	n.a.	0.05
Jamaica	2.38	n.a.	n.a.	0.22
Mexico	9.69	0.00	1.88	0.74
Nicaragua	0.33	n.a.	n.a.	n.a.
Panama	9.65	n.a.	n.a.	0.06
Paraguay	0.00	n.a.	n.a.	n.a.
Peru	1.53	0.00	1.65	0.20
St. Kitts and Nevis	3.62	n.a.	n.a.	n.a.
St. Lucia	3.60	n.a.	n.a.	n.a.
St. Vincent and the Grenadines	3.79	n.a.	n.a.	n.a.
Suriname	0.00	n.a.	n.a.	n.a.
Trinidad and Tobago	16.19	n.a.	17.71	n.a.
Uruguay	0.00	n.a.	0.05	0.00
Venezuela	23.58	3.59	3.56	0.01
MENA				
Algeria	15.91	2.89	16.08	0.00
Bahrain	36.25	10.44	9.67	n.a.
Djibouti	0.69	1.49	n.a.	n.a.
Egypt	40.16	11.35	13.87	0.32
Iran	32.17	14.66	28.62	0.09
Iraq	19.08	2.01	0.43	0.00
Jordan	21.36	15.49	1.90	n.a.
Kuwait	10.38	4.62	3.03	0.00
Lebanon	16.69	19.59	1.04	0.62
Libya	23.66	6.04	4.94	0.00
Mauritania	3.87	3.37	2.91	n.a.
Morocco	11.03	n.a.	0.19	1.68
Oman	16.26	2.27	9.33	n.a.
Qatar	14.30	3.26	5.38	0.00
Saudi Arabia	25.47	5.23	1.80	0.00
Sudan	12.83	n.a.	n.a.	n.a.
Syria	n.a.	n.a.	n.a.	n.a.
Tunisia	8.67	7.66	3.17	n.a.
United Arab Emirates	10.21	5.82	13.33	0.11
Yemen	29.54	5.99	6.23	n.a.
Sub-Saharan Africa				
Angola	5.42	0.64	0.11	0.00
Benin	2.36	9.98	n.a.	n.a.
Botswana	3.54	1.64	n.a.	1.53
Burkina Faso	1.50	4.30	n.a.	n.a.
Burundi	0.00	n.a.	n.a.	n.a.
Cameroon	13.62	12.76	0.37	n.a.
Cape Verde	0.00	10.23	n.a.	n.a.
Central African Republic	0.00	n.a.	n.a.	n.a.
Chad	0.00	0.06	n.a.	n.a.
Comoros	n.a.	n.a.	n.a.	n.a.

(Continued)

TABLE A.5 (CONCLUDED)

Country	Petroleum products	Electricity	Natural gas	Coal
Sub-Saharan Africa (concluded)				
Congo, Democratic Republic of the	0.00	6.57	0.02	0.46
Congo, Republic of	5.20	6.25	0.03	n.a.
Côte d'Ivoire	0.00	14.59	2.79	n.a.
Equatorial Guinea	6.56	n.a.	n.a.	n.a.
Eritrea	n.a.	n.a.	n.a.	n.a.
Ethiopia	4.06	7.89	n.a.	n.a.
Gabon	3.13	n.a.	0.32	n.a.
Gambia, The	0.00	n.a.	n.a.	n.a.
Ghana	3.03	15.50	n.a.	n.a.
Guinea	0.00	n.a.	n.a.	n.a.
Guinea-Bissau	0.00	n.a.	n.a.	n.a.
Kenya	2.44	0.66	n.a.	0.04
Lesotho	0.10	1.77	n.a.	n.a.
Liberia	0.07	n.a.	n.a.	n.a.
Madagascar	4.01	8.73	n.a.	n.a.
Malawi	0.58	6.83	n.a.	n.a.
Mali	0.82	4.24	n.a.	n.a.
Mauritius	0.00	n.a.	n.a.	n.a.
Mozambique	0.90	16.89	0.44	0.07
Namibia	0.45	1.82	n.a.	0.34
Niger	1.24	0.88	n.a.	n.a.
Nigeria	7.33	4.55	0.94	0.00
Rwanda	0.00	1.50	n.a.	n.a.
Senegal	0.00	11.22	0.04	1.02
Seychelles	0.00	n.a.	n.a.	n.a.
Sierra Leone	3.71	n.a.	n.a.	n.a.
South Africa	1.46	2.65	0.00	11.79
Swaziland	0.00	n.a.	n.a.	n.a.
São Tomé and Príncipe	1.70	n.a.	n.a.	n.a.
Tanzania	0.00	10.23	1.24	0.17
Togo	3.96	n.a.	n.a.	n.a.
Uganda	0.00	9.79	n.a.	n.a.
Zambia	0.00	22.07	n.a.	0.00
Zimbabwe	n.a.	48.22	n.a.	9.96
World	**3.12**	**0.77**	**1.61**	**3.04**

Sources: GIZ; IEA, *World Energy Outlook 2012;* IMF staff estimates; IMF, *World Economic Outlook;* OECD; World Bank.
Note: Values are rounded to the nearest one-hundredth percent; electricity subsidies are taken from 2009 for 31 countries, and natural gas data are taken from 2010 for four countries. World estimates are calculated as identified subsidies divided by global government revenues. n.a. = not applicable; CEE-CIS = Central and Eastern Europe and Commonwealth of Independent States; LAC = Latin America and the Caribbean; MENA = Middle East and North Africa.
[1]Owing to the lack of systematic cross-country data, the subsidy estimates do not take into account certain charges and taxes for energy that are often rationalized on environmental grounds. Principally, these include regional or country-level carbon pricing programs, road user charges (e.g., mileage-based tolls for trucks in Germany), and excise taxes on electricity consumption and vehicle sales. For example, the European Union Emissions Trading System imposes a carbon tax on certain greenhouse gases that are emitted by factories, power plants, and other installations in the system. The current prevailing prices, however, are only a small fraction of estimated damages, and only about half of emissions are covered under the scheme. To take another example, incorporating the road user charges for diesel vehicles in New Zealand would lower our posttax energy subsidy estimates for that country by US$0.8 billion.
[2]The estimate for Luxembourg reflects, to a large extent, cross-border sales of petroleum products to neighboring countries, with buyers attracted by lower tax rates.

Appendix B. Assessing the Environmental and Health Impacts of Energy Subsidy Reform

Ian Parry and Baoping Shang

This appendix describes the methodologies used to provide calculations of the impact of energy subsidy reform on CO_2 emissions, SO_2 emissions, and other local pollutants. Here we consider a scenario in which energy prices are raised to levels that would eliminate tax-inclusive subsidies for petroleum products, coal, natural gas, and electricity. For each product, after calculating the increase in prices that is needed to eliminate tax-inclusive subsidies, we estimate how much the quantity demanded decreases for the product. The decrease in quantity demanded is determined by the assumption for the elasticity of demand for the product. The details on how this is estimated for each product are described below.

PETROLEUM PRODUCTS

CO_2 Emissions

A price elasticity of -0.4 is assumed for gasoline, diesel, and kerosene (Parry, 2011). The reduction in CO_2 emissions is then calculated by multiplying the reduction in consumption by the CO_2 coefficient of 0.0023 tons per liter of gasoline. The CO_2 coefficient is assumed to be 16 percent higher for diesel and kerosene (Parry, 2011).

Local Pollution

The reduction (in percentage terms) in other local pollutants owing to fossil fuel combustion is approximated by the reduction in fuel consumption. Gasoline combustion produces only a small amount of SO_2, and thus the impact of petroleum subsidy removal on SO_2 is not estimated.

COAL

CO$_2$ Emissions

The reduction (in percent) in coal consumption is calculated by assuming a price elasticity of -0.2 (Energy Information Administration [EIA], 2012).[1] The reduction in CO$_2$ emissions resulting from the removal of coal subsidies is then estimated as the same reduction (in percent) in total CO$_2$ emissions from coal, based on Organisation for Economic Co-operation and Development (OECD) data.

SO$_2$ Emissions

This is estimated by using an SO$_2$ coefficient of 0.01 tons of SO$_2$ per short ton of coal (EIA, 2012; Environmental Protection Agency, 2012). Local pollution other than SO$_2$ from coal is considered minor.

NATURAL GAS

The reduction (in percent) in natural gas consumption is calculated by assuming a price elasticity of -0.3 (EIA, 2012). The reduction in CO$_2$ emissions is then estimated as the same percent reduction in total CO$_2$ emissions from natural gas, based on OECD data. As noted previously, the impact of natural gas use on local pollution is assumed to be relatively small.

ELECTRICITY

Electricity subsidies increase the consumption of coal, natural gas, and other generation fuels because of excess demand for electricity. However, for the following reasons, these effects on emissions are not quantified in this paper:

1. In some countries, part of the electricity subsidies is due to inefficiencies in the electricity sector. In other words, part of the problem is not that prices are too low but that costs are too high. Thus, successful subsidy reforms could reduce these inefficiencies without raising prices and suppressing demand.

2. Data limitations make it difficult to quantify the environmental impact of electricity subsidy removal. For example, price and cost data are limited, and there is a lack of information on the marginal energy source for electricity generation, which may be different from the average.

3. The environmental impact of price increases in fuel, coal, and natural gas as inputs for electricity generation is already incorporated in the calculations of these energy products. In addition, electricity subsidies are relatively small as a share of total posttax subsidies, so this omission is expected to have only a small impact on the overall estimates.

---

[1]An upward adjustment is made to the EIA estimate because it is generally viewed as being on the conservative side.

CAVEATS

The methods used here are used to provide some rough estimates of the magnitude of the impacts over the longer term. They have several limitations. For example, they do not take into account the substitution between different energy products and resulting offsetting effects (e.g., there could be some offsetting increase in emissions if subsidy removal raises the price of natural gas to coal).

References

Ajodhia, V., W. Mulder, and T. Slot, 2012, *Tariff Structures for Sustainable Electrification in Africa* (Arnhem, the Netherlands: KEMA).

Amavilah, V. H. S., 1999, "The Resellers' Regulated Demand Price of 93 Octane Petrol in Namibia Relative to OPEC Crude Oil Price, 1991," Discussion Paper No. 10, Ore Body Engineering Ltd.

Antmann, Pedro, 2009, "Reducing Technical and Non-Technical Losses in the Power Sector," Background Paper for the World Bank Group Energy Sector Strategy (Washington: World Bank).

Arze del Granado, Javier, David Coady, and Robert Gillingham, 2012, "The Unequal Benefits of Fuel Subsidies: A Review of Evidence for Developing Countries," *World Development*, Vol. 40 (November), pp. 2234–48.

Atiyas, Izak, and Mark Dutz, 2012, "Competition and Regulatory Reform in the Turkish Electricity Industry," in *Reforming Turkish Energy Markets* (New York: Springer).

Bacon, R., E. Ley, and M. Kojima, 2010, "Subsidies in the Energy Sector: An Overview," Background Paper for the World Bank Group Energy Sector Strategy, July (Washington: World Bank).

Baig, Taimur, Amine Mati, David Coady, and Joseph Ntamatungiro, 2007, "Domestic Petroleum Product Prices and Subsidies: Recent Developments and Reform Strategies," IMF Working Paper No. 07/71 (Washington: International Monetary Fund). Available via the Internet: http://www.imf.org/external/pubs/ft/wp/2007/wp0771.pdf

Bank of Namibia, 2005, *Quarterly Bulletin*, Vol. 14, No. 1 (March).

Beaton, Christopher, and Lucky Lontoh, 2010, "Lessons Learned from Indonesia's Attempts to Reform Fossil-Fuel Subsidies" (Winnipeg: International Institute for Sustainable Development).

Bernardo, Romeo L., Marie-Christine G. Tang, 2008, "The Political Economy of Reform during the Ramos Administration (1992–98)," Commission on Growth and Development Working Paper No. 39 (Washington: World Bank).

Besant-Jones, John E., 2006, "Reforming Power Markets in Developing Countries: What Have We Learned?" Energy and Mining Sector Board Discussion Paper No. 19 (Washington: World Bank).

Blaschke, W., and U. Lorenz, 2004, "Restructuring of Polish Hard Coal Industry in the Last Decade and Perspectives for the Next Decade," in *European Conference on Raw Building and Coal: New Perspectives* (Sarajevo: Wyd IP Svjetlost, d.d.). Available via the Internet: http://www.min-pan.krakow.pl/zaklady/zrynek/zasoby/04_04wb_ul_sarajewo.pdf

BP, 2012, *Statistical Review of World Energy: June 2012*. Available via the Internet: http://www.bp.com/statisticalreview

Breisinger, Clemens, Wilfried Engelke, and Oliver Ecker, 2011, "Petroleum Subsidies in Yemen: Leveraging Reform for Development," Policy Research Working Paper No. 5577 (Washington: World Bank).

Briceño-Garmendia, C., and M. Shkaratan, 2011a, "Kenya's Infrastructure: A Continental Perspective," Policy Research Working Paper No. 5596 (Washington: World Bank).

———, 2011b, "Power Tariffs: Caught between Costs Recovery and Affordability," Policy Research Working Paper No. 5904 (Washington: World Bank).

Burniaux, Jean-Marc, Jean Chateau, Romain Duval, and Stéphanie Jamet, 2009, "The Economics of Climate Change Mitigation: How to Build the Necessary Global Action in a Cost-Effective Manner," OECD Economics Department Working Papers No. 701 (Paris: Organisation for Economic Co-operation and Development).

Calderón, César, 2008, "Infrastructure and Growth in Africa," Policy Research Working Paper No. 4914 (Washington: World Bank).

Carreón-Rodriguez, Victor, Armando Jiménez San Vicente, and Juan Rosellón, 2003, "The Mexican Electricity Sector: Economic, Legal and Political Issues," Working paper (Stanford, CA: Program on Energy and Sustainable Development, Stanford University).

Clements, Benedict, Sanjeev Gupta, and Masahiro Nozaki, 2012, "What Happens to Social Spending in IMF-Supported Programs?" *Applied Economics*, Vol. 45, No. 28, pp. 4022–33.

Clements, Benedict, Hong-Sang Jung, and Sanjeev Gupta, 2007, "Real and Distributive Effects of Petroleum Price Liberalization: The Case of Indonesia," *The Developing Economies*, Vol. 45, No. 2, pp. 220–37.

Coady, David, Javier Arze del Granado, Luc Eyraud, Hui Jin, Vimal Thakoor, Anita Tuladhar, and Lilla Nemeth, 2012, "Automatic Fuel Pricing Mechanisms with Price Smoothing: Design, Implementation, and Fiscal Implications," Technical Notes and Manuals No. 12/03 (Washington: International Monetary Fund). Available via the Internet: http://www.imf.org/external/pubs/ft/tnm/2012/tnm1203.pdf

Coady, David, Moataz El-Said, Robert Gillingham, Kangni Kpodar, Paulo Medas, and David Newhouse, 2006, "The Magnitude and Distribution of Fuel Subsidies: Evidence from Bolivia, Ghana, Jordan, Mali, and Sri Lanka," IMF Working Paper No. 06/247 (Washington: International Monetary Fund).

Coady, David, Robert Gillingham, Rolando Ossowski, John Piotrowski, Shamsuddin Tareq, and Justin Tyson, 2010, "Petroleum Product Subsidies: Costly, Inequitable, and Rising," IMF Staff Position Note No. 10/05 (Washington: International Monetary Fund). Available via the Internet: http://www.imf.org/external/pubs/ft/spn/2010/spn1005.pdf

Coady, David, and David Newhouse, 2006, "Evaluating the Distribution of the Real Income Effects of Increases in Petroleum Product Prices in Ghana," in *Analyzing the Distributional Impacts of Reforms: Operational Experience in Implementing Poverty and Social Impact Analysis*, ed. by A. Coudouel, A. Dani, and S. Paternostro (Washington: World Bank).

Competition Tribunal of South Africa, 2006, "Uhambo Merger Findings," Case No: 101/LM/Dec04.

Deutsche Gesellschaft für Internationale Zusammenarbeit, 2009, *International Fuel Prices 2009* (Bonn and Berlin, 6th ed.).

Dick, Herman, Sanjeev Gupta, David Vincent, and Herbert Voight, 1984, "The Impact of Oil Price Increases on Four Oil-Poor Developing Countries: A Comparative Analysis," *Energy Economics*, Vol. 6 (January), pp. 59–70.

Ebert, Sebastian, Gerhard P. Metschies, Dominik Schmid, and Armin Wagner, 2009, *International Fuel Prices 2009* (Eschborn: Deutsche Gesellschaft für Technische Zusammenarbeit, 6th ed.).

Ellis, Jennifer, 2010, "The Effects of Fossil-Fuel Subsidy Reform: A Review of Modelling and Empirical Studies," in *The Untold Billions: Fossil-Fuel Subsidies, Their Impacts and the Path to Reform* (Geneva: Global Subsidies Initiative).

Energy Information Administration (EIA), 2012, "Fuel Competition in Power Generation and Elasticities of Substitution" (Washington, D.C.: U.S. Department of Energy).

Environmental Protection Agency, 2012, "Quarterly Emissions Tracking," U.S. Environmental Protection Agency. Available via the Internet: http://www.epa.gov/airmarkt/quarterly tracking.html

Escribano, Alvaro, J. Luis Guasch, and Jorge Pena, 2008, "A Robust Assessment of the Impact of Infrastructure on African Firms' Productivity," Africa Infrastructure Country, Diagnostic Working Paper (Washington: World Bank).

Fernandez, Luisa, and Rosechin Olfindo, 2011, "Overview of the Philippines' Conditional Cash Transfer Program," Social Protection Note No. 2, May (Washington: World Bank).

Fiszbein, Ariel, and Norbert Schady, 2009, *Conditional Cash Transfers: Reducing Present and Future Poverty*, Policy Research Report (Washington: World Bank).

Fofana, Ismaél, Margaret Chitiga, and Ramos Mabugu, 2009, "Oil Prices and the South African Economy: A Macro-Meso-Micro Analysis," *Energy Policy*, Vol. 37 (December), pp. 5509–18.

Foster, V., and C. Briceño-Garmendia, 2010, *Africa's Infrastructure: A Time for Transformation* (Washington: World Bank).

Foster, Vivien, and Jevgenijs Steinbuks, 2008, "Paying the Price for Unreliable Power Supplies: In-House Generation of Electricity by Firms in Africa," Policy Research Working Paper No. 4913 (Washington: World Bank).

Garcia, Morito, and Charity M. T. Moore, 2012, *The Cash Dividend: The Rise of Cash Transfers in Sub-Saharan Africa*, (Washington: World Bank).

Gelb, Alan, and associates, 1988, *Oil Windfalls: Blessing or Curse?* (New York: Oxford University Press).

Giambiagi, F., and M. Moreira, 1999, "A Economia Brasileira nos Anos 90" (Rio de Janeiro: Banco Nacional de Desenvolvimento Econômico e Social).

Global Subsidies Initiative, 2010, "Strategies for Reforming Fossil-Fuel Subsidies: Practical Lessons from Ghana, France and Senegal," in *The Untold Billions: Fossil-Fuel Subsidies, Their Impacts and the Path to Reform* (Winnipeg: International Institute for Sustainable Development).

Graham, Carol, 1998, *Private Markets for Public Goods: Raising the Stakes in Economic Reform* (Washington: World Bank).

Grosh, Margaret, Carlo del Ninno, Emil Tesliuc, and Azedine Ouerghi, 2008, *For Promotion and Protection: The Design and Implementation of Effective Safety Nets* (Washington: World Bank).

Gupta, Sanjeev, 1983, "India and the Second OPEC Oil Shock—An Economy-Wide Analysis," *Review of World Economics*, Vol. 119, No. 1, pp. 122–37.

Gupta, Sanjeev, Benedict Clements, Kevin Fletcher, and Isabel Inchauste, 2004, "Issues in Domestic Petroleum Pricing in Oil-Producing Countries," in *Fiscal Policy Formulation and Implementation in Oil-Producing Countries,* ed. by J. Davis, R. Ossowski, and A. Fedelino (Washington: International Monetary Fund).

Gupta, Sanjeev, Marijn Verhoeven, Robert Gillingham, Christian Schiller, Ali Mansoor, and Juan Pablo Cordoba, 2000, *Equity and Efficiency in the Reform of Price Subsidies: A Guide for Policymakers* (Washington: International Monetary Fund).

Heggie, Ian G., and Piers Vickers, 1998, "Commercial Management and Financing of Roads," World Bank Technical Paper No. 409 (Washington: World Bank). Available via the Internet: http://documents.worldbank.org/curated/en/1998/05/441617/commercial-management-financing-roads

Institute for Fiscal Studies, 2012, "Tax and Benefit Tables" (London). Available via the Internet: http://www.ifs.org.uk/fiscalFacts/taxTables

International Energy Agency, 2010, "Energy Policies of IEA Countries: Turkey 2009 Review" (Paris).

———, 2011a, "Development in Energy Subsidies," in the *World Energy Outlook* (Paris).

———, 2011b, "Fossil-Fuel Subsidies—Methodology and Assumptions," *World Energy Outlook*. Available via the Internet: http://www.iea.org/publications/worldenergyoutlook/resources/energysubsidies/methodologyforcalculatingsubsidies

———, 2011c, *World Energy Outlook* (Paris).

International Finance Corporation, 2012, *From Gap to Opportunity: Business Models for Scaling up Energy Access* (Washington).

International Monetary Fund, 2001, "Ghana," Country Report No. 01/141 (Washington).

———, 2005a, "Ghana," Country Report No. 05/292 (Washington).

———, 2005b, "Power Sector Reform in the Philippines," in *Philippines: Selected Issues Paper* (Washington).

———, 2008, "Food and Fuel Prices—Recent Developments, Macroeconomic Impact, and Policy Responses" (Washington). Available via the Internet: http://www.imf.org/external/np/pp/eng/2008/063008.pdf

————, 2011a, "Ghana," Country Report No. 11/128 (Washington).

————, 2011b, *Regional Economic Outlook: Middle East and Central Asia*, World Economic and Financial Surveys (Washington).

————, 2012a, "Fiscal Transparency, Accountability, and Risk," IMF Policy Paper (Washington). Available via the Internet: https://www.imf.org/external/np/pp/eng/2012/080712.pdf

————, 2012b, "Ghana," Country Report No. 12/36 (Washington).

————, 2012c, "Ghana," Country Report No. 12/201 (Washington).

————, 2012d, "Managing Global Growth Risks and Commodity Price Shocks: Vulnerabilities and Policy Challenges for Low-Income Countries" (Washington). Available via the Internet: http://www.imf.org/external/np/pp/eng/2011/092111.pdf

————, 2012e, *Regional Economic Outlook: Sub-Saharan Africa: Sustaining Growth amid Global Uncertainty*, World Economic and Financial Surveys (Washington).

————, forthcoming, "Pricing Energy for Environmental Damages: Putting Principle into Practice" (Washington).

Komives, Kristin, Todd Johnson, Jonathan Halpern, Jose Luis Aburto, and John Scott, 2009, "Residential Electricity Subsidies in Mexico: Exploring Options for Reform and for Enhancing the Impact on the Poor," Working Paper No. 160 (Washington: World Bank).

Koplow, Doug, 2009, "Measuring Energy Subsidies Using the Price-Gap Approach: What Does It Leave out?" IISD Trade, Investment and Climate Change Series (Winnipeg: International Institute for Sustainable Development). Available via the Internet: http://www.iisd.org/publications/pub.aspx?pno=1165

Kumar, Manmohan S., and Jaejoon Woo, 2010, "Public Debt and Growth," IMF Working Paper No. 10/174 (Washington: International Monetary Fund). Available via the Internet: http://www.imf.org/external/pubs/ft/wp/2010/wp10174.pdf

Kumar, Manmohan S., Anthony A. Kolb, Sumila Gulyani, and Vahram Avenesyan, 2011, "Utility Pricing and the Poor: Lessons from Armenia," Technical Paper No. 497 (Washington: World Bank).

Lampietti, Julian, Anthony A. Kolb, Sumila Gulyani, and Vahram Avenesyan, 2011, "Utility Pricing and the Poor: Lessons from Armenia," World Bank Technical Papers No. 497 (Washington: World Bank).

Larrain, Felipe B., 2010, "Nuevos Mecanismos de Protección ante Variaciones de Precios de Combustibles," Presentation, Chilean Ministry of Finance. Available via the Internet: http://www.chiletransporte.cl/3w/images/Documentos/Sipco.pdf

Lofgren, Hans, 1995, "Macro and Micro Effects of Subsidy Cuts: A Short Run CGE Analysis for Egypt," TMD Discussion Paper No. 5 (Washington: International Food Policy Research Institute).

Márquez, Miguel, 2000, "El Fondo de Estabilización de Precios del Petróleo (FEPP) y el Mercado de los derivados en Chile," División de Recursos Naturales e Infraestructura, ECLAC. Available via the Internet: http://www.eclac.org/publicaciones/xml/9/5789/Lcl1452e.pdf

Mota, Raffaella, 2003, "The Restructuring and Privatisation of Electricity Distribution and Supply Business in Brazil: A Social Cost-Benefit Analysis," Working Paper No. WP 0309 (Cambridge: University of Cambridge, Department of Applied Economics).

Mourougane, Annabelle, 2010, "Phasing out Energy Subsidies in Indonesia," OECD Economics Department Working Paper No. 808 (Paris: Organisation for Economic Co-operation and Development).

National Research Council, 2009, "Hidden Costs of Energy: Unpriced Consequences of Energy Production and Use," Committee on Health, Environmental, Other External Costs and Benefits of Energy Production and Consumption (Washington: National Academies Press).

Nixson, Frederick, and Bernard Walters, 2005, "Utilities' Pricing and the Poor: The Case of Armenia," UNDP Armenia White Paper (New York: United Nations Development Program).

Nordhaus, William, 2011, "Estimates of the Social Cost of Carbon: Background and Results from the RICE-2011 Model," NBER Working Paper No. 17540 (Cambridge: National Bureau of Economic Research).

Okigbo, Patrick O., and Dili Enekebe, 2011, "Nigeria: Fuel Subsidy Removal—Achieving the Optimal Solution," Nextier Policy Brief, December 15 (Washington: Nextier).

Okonjo-Iweala, Ngozi, 2011, "Brief on Fuel Subsidy," December 6 (Nigeria: Federal Ministry of Finance).

Organisation for Economic Co-operation and Development, 2004, *Mexico: Progress in Implementing Regulatory Reform* (Paris).

———, 2009, "The Economics of Climate Change Mitigation Policies and Options for Global Action beyond 2012" (Paris).

———, 2012a, "Inventory of Estimated Budgetary Support and Tax Expenditures for Fossil Fuels" (Paris). Available via the Internet: http://www.oecd.org/tad/environmentandtrade/inventoryofestimatedbudgetarysupportandtaxexpendituresforfossilfuels.htm

———, 2012b, "Mortality Risk Valuation in Environment, Health and Transport Policies" (Paris).

———, 2013, "Chile: Inventory of Estimated Budgetary Support and Tax Expenditures for Fossil Fuels," Overview Note, OECD-IEA Fossil Fuel Subsidies and Other Support. Available via the Internet: http://www.oecd.org/site/tadffss/Chileoverviewfossilfuelsupport2013.pdf

O'Ryan, Raul, Sebastian Miller, Jorge Rogat, and Carlos de Miguel, 2003, "The Impact of Removing Energy Subsidies in Chile," in *Energy Subsidies: Lessons Learned in Assessing Their Impact and Designing Policy Reforms* (Geneva: United Nations Environment Program). Available via the Internet: http://www.unep.ch/etb/publications/energySubsidies/Energysubreport.pdf

Parry, Ian W. H., 2011, "How Much Should Highway Fuels Be Taxed?" in *U.S. Energy Tax Policy*, ed. by Gilbert E. Metcalf (Cambridge: Cambridge University Press).

Parry, Ian W. H., and Kenneth A. Small, 2005, "Does Britain or the United States Have the Right Gasoline Tax?" *American Economic Review*, Vol. 95, No. 4, pp. 1276–89.

Parry, Ian W. H., and Jon Strand, 2011, "International Fuel Tax Assessment: An Application to Chile," IMF Working Paper No. 11/168 (Washington: International Monetary Fund). Available via the Internet: http://www.imf.org/external/pubs/ft/wp/2011/wp11168.pdf

Philippines, Department of Energy, 2006, "8th EPIRA Implementation Status Report" (Manila).

Ranganathan, R., and Vivien Foster, 2012, "Uganda's Infrastructure: A Continental Perspective," Policy Research Working Paper No. 5963 (Washington: World Bank).

Sargsyan, Gevorg, Ani Balabanyan, and Denzel Hankinson, 2006, "From Crisis to Stability in the Armenian Power Sector: Lessons Learned from Armenia's Energy Reform Experience," Working Paper No. 74 (Washington: World Bank).

Stern, Nicholas, 2006, *Stern Review on the Economics of Climate Change* (London: Her Majesty's Treasury).

Sterner, Thomas, ed., 2012, *Fuel Taxes and the Poor: The Distributional Effects of Gasoline Taxation and Their Implications for Climate Policy* (Washington: RFF Press).

Suwala, Wojciech, 2010, "Lessons Learned from the Restructuring of Poland's Coal-Mining Industry" (Geneva: Global Subsidies Initiative). Available via the Internet: http://www.iisd.org/gsi/sites/default/files/poland_casestudy_ffs.pdf

Uganda, Ministry of Energy and Mineral Development, 2012a, *Energy and Mineral Sector Performance Report, 2008–2011* (Kampala).

———, 2012b, *Performance of the Uganda Power Sector* (Kampala).

United Nations Environment Program and the International Energy Agency, 2002, "Reforming Energy Subsidies: An Explanatory Summary of the Issues and Challenges in Removing or Modifying Subsidies on Energy that Undermine the Pursuit of Sustainable Development" (Paris).

United Nations Environment Program, 2008, "Reforming Energy Subsidies: Opportunities to Contribute to the Climate Change Agenda," Division of Technology, Industry and Economics (Paris).

United States, Interagency Working Group on Social Cost of Carbon, 2013, "Technical Support Document: Technical Update of the Social Cost of Carbon for Regulatory Impact Analysis Under Executive Order 12866." Available via the Internet: http://www.whitehouse.gov/sites/default/files/omb/inforeg/social_cost_of_carbon_for_ria_2013_update.pdf

Vagliasindi, Maria, 2013, "Implementing Energy Subsidy Reforms: Evidence from Developing Countries" (Washington: World Bank). Available via the Internet: https://openknowledge.worldbank.org/handle/10986/11965

Velody, Mark, Michael J. G. Cain, and Michael Philips, 2003, "Energy Reform and Social Protection in Armenia," in *A Regional Review of Social Safety Net Approaches: In Support of Energy Sector Reform* (Washington: United States Agency for International Development).

von Moltke, Anja, Colin McKee, and Trevor Morgan, 2004, *Energy Subsidies: Lessons Learned in Assessing Their Impact and Designing Policy Reforms* (Sheffield: Greenleaf Publishing).

World Bank, 2008, *Philippines Quarterly Update* (Washington). Available via the Internet: http://documents.worldbank.org/curated/en/2008/11/11962976/philippines-quarterly-update

———, 2009, "Philippines—First Development Policy Loan Project." Available via the Internet: http://documents.worldbank.org/curated/en/2009/09/11415139/philippines-first-development-policy-loan-project

———, 2010a, "Project Appraisal Document for the Electricity Expansion Project," Report No. 54147-KE, May (Washington).

———, 2010b, "Subsidies in the Energy Sector: An Overview," Background paper for the World Bank Group Energy Sector Strategy, July. Available via the Internet: http://siteresources.worldbank.org/EXTESC/Resources/Subsidy_background_paper.pdf

———, 2011a, "Uganda: Strategic Review of the Electricity Sector," Draft policy note (Washington).

———, 2011b, "Proposed Credit to the Republic of Uganda for an Electricity Sector Development Project," Report No. 59310-UG, May (Washington).

———, 2012a, *Improving the Targeting of Social Programs in Ghana*, ed. by Q. Wendon (Washington).

———, 2012b, "Proposed Credit in the Amount of SDR 32.3 million (US$50 million equivalent) to the Republic of Niger for a First Shared Growth Credit," Report No. 69131-NE, May 31 (Washington).

Zhang, Fan, 2011, "Distributional Impact Analysis of the Energy Price Reform in Turkey," Policy Research Working Paper No. 5831 (Washington: World Bank).

Contributors

Trevor Alleyne, Advisor, African Department
Andreas Bauer, Assistant Director, Strategy, Policy, and Review Department
Benedict Clements, Division Chief, Fiscal Affairs Department
David Coady, Deputy Division Chief, Fiscal Affairs Department
Antonio David, Economist, Institute for Capacity Development
Ozgur Demirkol, Senior Economist, Middle East and Central Asia Department
Allan Dizioli, Economist, Fiscal Affairs Department
Stefania Fabrizio, Deputy Division Chief, Fiscal Affairs Department
Katja Funke, Economist, Fiscal Affairs Department
Sanjeev Gupta, Deputy Director, Fiscal Affairs Department
Farayi Gwenhamo, Economist, African Department
Mumtaz Hussain, Senior Economist, African Department
Christian Josz, Deputy Division Chief, African Department
Alvar Kangur, Economist, Fiscal Affairs Department
Javier Kapsoli, Economist, Fiscal Affairs Department
Kangni Kpodar, Senior Economist, Fiscal Affairs Department
Clara Mira, Economist, African Department
Luc Moers, Senior Economist, Middle East and Central Asia Department
Masahiro Nozaki, Senior Economist, Fiscal Affairs Department
Anton Op de Beke, Resident Representative, African Department
Dragana Ostojic, Economist, Middle East and Central Asia Department
Ian Parry, Technical Assistance Advisor, Fiscal Affairs Department
Edgardo Ruggiero, Senior Economist, African Department
Carlo Sdralevich, Deputy Division Chief, Middle East and Central Asia Department
Louis Sears, Research Assistant, Fiscal Affairs Department
Baoping Shang, Economist, Fiscal Affairs Department
Sukhwinder Singh, Assistant to the Director, African Department
Mauricio Soto, Economist, Fiscal Affairs Department
Vimal Thakoor, Economist, African Department
Genevieve Verdier, Senior Economist, African Department
Mauricio Villafuerte, Deputy Division Chief, African Department
Younes Zouhar, Senior Economist, Middle East and Central Asia Department